REVISED REPRINT

MILLER'S

GUIDE
TO THE
DISSECTION
OF THE DOG

HOWARD E. EVANS, Ph.D.

Professor of Veterinary Anatomy
Department of Anatomy, New York
State Veterinary College, Cornell University

AND

ALEXANDER deLAHUNTA, D.V.M., Ph.D.

Assoc. Professor of Veterinary Anatomy
Department of Anatomy, New York
State Veterinary College, Cornell University

W. B. SAUNDERS COMPANY
PHILADELPHIA · LONDON · TORONTO

W. B. Saunders Company: West Washington Square
                          Philadelphia, Pa. 19105

                          12 Dyott Street
                          London, WC1A   1DB

                          833 Oxford Street
                          Toronto 18, Ontario

Listed here is the latest translated edition of this book together with the
language of the translation and the publisher.

Spanish (*1st Edition*) — Nueva Editorial Interamericana, S.A. de C.V.,
           Mexico City, Mexico

Miller's Guide to the Dissection of the Dog — Revised Reprint   ISBN 0-7216-3443-5

Last digit is the print number: 9   8   7   6   5   4   3   2

PREFACE

Dr. Malcolm E. Miller, D.V.M., Ph.D., who died in 1960, was Professor and Head of the Department of Anatomy in the New York State Veterinary College at Cornell University. His *Guide to the Dissection of the Dog,* first published in 1947, was revised in 1948 and 1952. Corrected reprints appeared in 1955, 1958 and 1962. The present *Guide* incorporates material from previous editions, eliminates some detail, alters several dissection procedures and adds new descriptive material, particularly on the nervous system.

The purpose of the *Guide* is to facilitate a careful dissection of the dog in order to learn basic mammalian structure. The description is based upon the dissection of embalmed, arterially injected adult dogs. The anatomical terms used are those which appear in the Nomina Anatomica Veterinaria 1968. A more detailed illustrated consideration of the structures dissected can be found in Miller, Christensen and Evans, *Anatomy of the Dog,* published by the W. B. Saunders Company, Philadelphia, 1964.

We are grateful to the many students and colleagues in various veterinary colleges for their helpful suggestions. The willingness of members of the American Association of Veterinary Anatomists and the World Association of Veterinary Anatomists to consider questions raised and to offer advice is appreciated. Much stimulation and help has come from our immediate associates in the Department of Anatomy, particularly Dr. R. E. Habel, who made many corrections and suggestions for improvement. Marion Newson has applied her talent as a medical illustrator to clarify the structures to be seen by preparing 53 new illustrations to augment those of Pat Barrow and her own which have appeared previously. Mrs.

Mary Miller Ewing, who assisted in the preparation of earlier editions, has encouraged our efforts to update the *Guide*.

HOWARD E. EVANS

ALEXANDER DELAHUNTA

Ithaca, New York

CONTENTS

INTRODUCTION... 1
    Medical Etymology and Anatomical Nomenclature................ 1
    Directional Terms 2
    Dissection... 5

THE SKELETAL AND MUSCULAR SYSTEMS........................ 6
    Bones of the Thoracic Limb................................... 6
    Muscles of the Thoracic Limb 17
    Joints of the Thoracic Limb................................. 45
    Bones of the Pelvic Limb.................................... 46
    Muscles of the Pelvic Limb 56
    Joints of the Pelvic Limb................................... 79
    Bones of the Vertebral Column.............................. 84
    Muscles of the Trunk 90

THE NECK, THORAX AND THORACIC LIMB.......................... 102
    Vessels and Nerves of the Neck 102
    The Thorax... 105
    Introduction to the Autonomic Nervous System 118
    Heart and Pericardium..................................... 123
    Vessels and Nerves of the Thoracic Limb................... 127

THE ABDOMEN, PELVIS AND PELVIC LIMB......................... 145
    Vessels and Nerves of the Ventral and Lateral Parts of the
        Abdominal Wall.. 145

Abdominal Viscera.. 152
Pelvic Viscera, Vessels and Nerves 178
Vessels and Nerves of the Pelvic Limb 189

THE HEAD.. 210
The Skull... 210
Structures of the Head.. 224

THE NERVOUS SYSTEM 257
Cerebral Meninges.. 257
Arteries 258
Veins.. 259
Brain.. 259
Spinal Cord... 271

INTRODUCTION

Anatomy is the study of structure. Physiology is the study of function. Structure and function are inseparable as the foundation of the science and art of medicine. One must know the parts before one can appreciate how they work. **Gross anatomy,** the study of structures that can be dissected and observed with the unaided eye or with a hand lens, forms the subject matter of this guide.

The anatomy of one part in relation to other parts of the body is **topographic anatomy.** The practical application of such knowledge in relation to the diagnosis and treatment of pathological or surgical conditions is **applied anatomy.** The study of structures too small to be seen without a microscope is **microscopic anatomy.** Examination of structure in even greater detail is possible with an electron microscope and constitutes **ultrastructural anatomy.** When an animal becomes diseased or its organs function improperly, its deviation from the normal is studied as **morbid anatomy** or **pathology.** The study of the development of the individual from the fertilized egg to birth is **embryology,** from the zygote to the adult, **developmental anatomy. Teratology** is the study of abnormal development.

MEDICAL ETYMOLOGY AND ANATOMICAL NOMENCLATURE

The student of anatomy is confronted with an array of unfamiliar terms and names of anatomical structures. A better understanding of the language of anatomy helps make its study more intelligible and interesting. For the publication of scientific papers and communication with colleagues, the mastery of anatomical terminology is a necessity. To assure knowledge of basic anatomical terms, a medical dictionary should be kept readily accessible and consulted frequently. It is very important to learn the spelling, pronunciation and meaning of all new terms encountered. Vertebrate structures are numerous and in many instances common names are not available or are so vague as to be meaningless. One soon realizes why it is desirable to have an international glossary of terms which can be understood by scientists in all countries. The learning of a medical vocabulary can be aided by the mastery of Greek and Latin roots and affixes.

Our present medical vocabulary has a history of over 2000 years and reflects the influences of the world's languages. The early writings in anatomy and medicine were almost entirely in Latin and as a consequence the majority of anatomical terms stem from this classical language. Latin terms are commonly translated into the vernacular of the person using them. Thus the Latin *hepar* becomes the English *liver,* the French *foie,* the Spanish *higado* and the German *Leber.*

Although anatomical terminology has been rather uniform, differences in terms have arisen between the different fields and different countries. In 1895 a group consisting mainly of German anatomists proposed a standard list of terms from those currently in use throughout the world. This list, known as the Basle Nomina Anatomica (BNA), was not adopted internationally but forms the basis for the present Nomina Anatomica (NA), which was adopted by the International Congress of Anatomists in Paris in 1955. Of the 5640 standard terms, over 80 per cent are continued from the BNA. The latest revision of the Nomina Anatomica is the third edition, approved in 1965 and published by the Excerpta Medica Foundation in 1966.

The guiding principles as stated in the Nomina Anatomica are to make no changes for purely pedantic or etymological reasons; to keep terms short and simple; to give similar names to closely related structures; to discard eponyms; to use terms with some informative or descriptive value; to arrange differentiating adjectives as opposites (for example: superficialis and profundus); to adopt single terms as a rule and to allow official alternatives only as exceptions; to resist naming every minute structure ever discovered or described; and to use Latin for all terms. Considerable thought and time has gone into anatomical terminology and committees are actively engaged in improving it. The International Committee on Veterinary Anatomical Nomenclature, appointed by the World Association of Veterinary Anatomists, published **Nomina Anatomica Veterinaria** for domestic animals in 1968. These Nomina serve as the basis for the nomenclature used in this guide.

DIRECTIONAL TERMS

An understanding of the following planes, positions and directions relative to the animal body or its parts is necessary in order to follow the procedures for dissection (Fig. 1).

PLANE: a surface, real or imaginary, along which any two points can be connected by a straight line.

Median plane: divides the head, body or limb longitudinally into equal right and left halves.

Sagittal plane: passes through the head, body or limb parallel to the median plane.

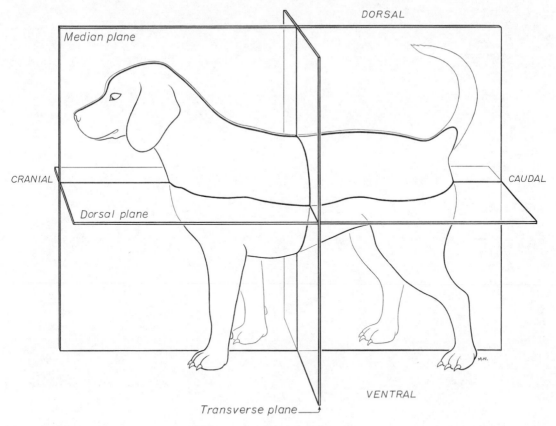

Figure 1. Directional terms.

Transverse plane: cuts across the head, body or limb at a right angle to its long axis or across the long axis of an organ or a part.

Dorsal plane: runs at right angles to the median and transverse planes and thus divides the body or head into dorsal and ventral portions.

DORSAL: toward or relatively near to the top of the head, back of the neck, trunk or tail; on the limbs it applies to the upper or front surface of the carpus, tarsus, metapodium and digits.

VENTRAL: toward or relatively near to the underside of the head and body.

MEDIAL: toward or relatively near to the median plane.

LATERAL: away from or relatively farther from the median plane.

CRANIAL: toward or relatively near to the head; on the limbs it applies proximal to the carpus and tarsus.

CAUDAL: toward or relatively near to the tail; on the limbs it applies proximal to the carpus and tarsus.

ROSTRAL: toward or relatively near to the nose; applies to the head only.

The adjectives for directional terms may be modified to serve as adverbs by replacing the ending "al" with the ending "ally," as in *dorsally.*

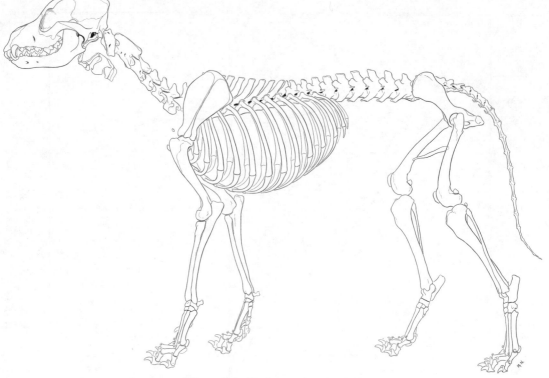

Figure 2.   *Skeleton of male dog.*

Certain terms whose meanings are more restricted are used in the description of organs and appendages.

INTERNAL OR INNER: close to, or in the direction of, the center of a hollow organ.

EXTERNAL OR OUTER: away from the center of a hollow organ.

SUPERFICIAL: relatively near to the surface of the body, or to the surface of a solid organ.

DEEP: relatively near to the center of the body or the center of a solid organ.

PROXIMAL: relatively near to the main mass or origin; in the appendages, the attached end.

DISTAL: away from the main mass or origin; in the appendages, the free end.

RADIAL: on that side of the forearm in which the radius is located.

ULNAR: on that side of the forearm in which the ulna is located.

TIBIAL AND FIBULAR: on the corresponding sides of the leg, the tibial side being medial, the fibular lateral.

The paw is that part of the thoracic or pelvic limb, in the dog or similar species, distal to the radius and ulna or to the tibia and fibula. The hand (manus) and foot (pes) of man are homologous with the forepaw and hind paw respectively.

PALMAR: the aspect of the forepaw on which the pads are located—the surface which faces the ground.

PLANTAR: the aspect of the hind paw on which the pads are located — the surface which faces the ground. The opposite surface of both forepaw and hind paw is known as the dorsal surface.

AXIS: the central line of the body or any of its parts.

AXIAL, ABAXIAL: of, pertaining to, or relative to the axis. In referring to the digits, the functional axis of the limb passes between the third and fourth digits. The axial surface of the digit faces the axis, the abaxial surface faces away from the axis.

The following terms apply to the various basic movements of the parts of the body.

FLEXION: the movement of one bone in relation to another in such a manner that the angle formed at their joint is reduced: the limb is retracted or folded; the digit is bent; the back is arched.

EXTENSION: the movement of one bone upon another is such that the angle formed at their joint increases: the limb reaches out or is extended; the digit is straightened; the back is straightened. Extension beyond 180 degrees is overextension.

ABDUCTION: the movement of a part away from the median plane.

ADDUCTION: the movement toward the median plane.

CIRCUMDUCTION: the movement of a part when outlining the surface of a cone (for example: the arm extended drawing a circle).

ROTATION: the movement of a part around its long axis (for example: the action of the radius when using a screwdriver).

DISSECTION

The dog provided as a specimen for dissection has been humanely prepared by injection of pentobarbital for anesthesia via the cephalic vein and by exsanguination through a cannula inserted in the common carotid artery. This procedure allows the pumping action of the heart to empty the blood vessels prior to the injection of embalming fluid consisting of 8% formalin and 2% phenol in aqueous solution. It is injected under 5 lb. of pressure over a period of approximately 30 minutes. After embalming, the arteries are injected with red latex, also through the common carotid artery, using a 50 cc. hand syringe. The prepared specimen represents a considerable investment of materials and labor and should be cared for accordingly. A well kept specimen facilitates dissection and study throughout the term. Gauze, 2% Phenoxetol or phenol and plastic sheeting are provided to moisten and wrap the paws and head and to cover the entire specimen between dissection periods. Refrigeration is helpful for storage but not essential.

There are certain principles and procedures which are generally accepted as aids in the learning of anatomy. The purpose of the dissection is to gain a clear understanding of the normal structures of the body.

THE SKELETAL
AND MUSCULAR SYSTEMS

The **appendicular skeleton** includes the bones of the thoracic girdle and forelimbs and the pelvic girdle and hind limbs.

THORACIC LIMB (forelimb)		PELVIC LIMB (hind limb)	
Thoracic girdle	Scapula Clavicle	Pelvic girdle	Ilium Ischium Pubis
Arm or Brachium	Humerus	Thigh	Femur
Forearm or Antebrachium	Radius Ulna	Leg or Crus	Tibia Fibula
Forepaw or Manus	Carpal bones Metacarpal bones Phalanges	Hind paw or Pes	Tarsal bones Metatarsal bones Phalanges

BONES OF THE THORACIC LIMB

The **thoracic girdle** consists of paired scapulae and clavicles. The scapula is large, whereas the clavicle is often reduced or absent. The dog's clavicle (Fig. 10) is a small oval plate located cranial to the shoulder in the brachiocephalicus muscle. The clavicle is one of the first bones to show a center of ossification in the fetal dog but in the adult it is partly cartilaginous.

SCAPULA

The scapula (Figs. 3, 4), a flat, roughly triangular bone, possesses two surfaces, three borders and three angles. The articular end forms the **glenoid cavity** and the constricted part which unites with the expanded blade is referred to as the **neck.**

The lateral surface (Fig. 3) of the scapula is divided into two nearly

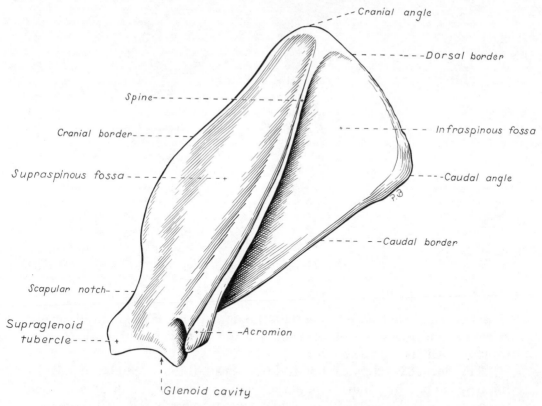

Figure 3. *Scapula, lateral surface.*

equal fossae by a shelf of bone, the **spine** of the scapula. The spine is the most prominent feature of the bone. It begins at the dorsal border as a thick, low ridge and becomes thinner and wider toward the neck. In all breeds the free border is slightly thickened, and in some it is everted caudally. The distal end is a truncated process, the **acromion,** where part of the deltoideus muscle arises. On a continuation of the spine proximally the omotransversarius attaches. The remaining part of the spine provides a place for insertion of the trapezius and for origin of that part of the deltoideus which does not arise from the acromion.

The **supraspinous fossa** is the entire surface cranial to the spine of the scapula. The supraspinatus arises from all but the distal part of this fossa.

The **infraspinous fossa,** caudal to the spine, is triangular with the apex at the neck. The infraspinatus arises from the infraspinous fossa.

The medial or costal surface presents two areas. A small dorsocranial rectangular area, the **serrated face,** serves as insertion for the powerful serratus ventralis. The large remaining part of the costal surface is the **subscapular fossa** which is nearly flat and usually presents three straight muscular lines that converge distally. The subscapularis arises from the whole subscapular fossa.

The **cranial border** of the scapula is thin. Near the ventral angle the border is concave as it enters into the formation of the neck. The notch

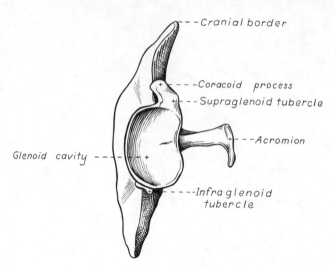

Figure 4.   Scapula, ventral angle.

thus formed is the **scapular notch.** The dorsal end of the cranial border thickens, and, without definite demarcation at the cranial angle, is continuous with the dorsal border.

The **dorsal border** extends from the cranial to the caudal angles. In life it is capped by a narrow band of cartilage but in the dried specimen the cartilage is destroyed by ordinary preparation methods. The rhomboideus attaches to this border.

The thick **caudal border** bears just dorsal to the ventral angle the **infraglenoid tubercle** from which arise the teres minor and the long head of the triceps. The middle third of the caudal border of the scapula is broad and smooth; part of the subscapularis and long head of the triceps arise from it. Somewhat less than the dorsal third of the caudal border is thick and gives rise to the teres major.

The **ventral angle** forms the expanded distal end of the scapula. The adjacent constricted part, the neck, is the segment of the scapula ventral to the spine and dorsal to the expanded part of the bone which forms the glenoid cavity. Clinically, the ventral angle is by far the most important part of the scapula, since it enters into the formation of the shoulder. The glenoid cavity articulates with the head of the humerus. Observe the shallowness of the cavity.

The **supraglenoid tubercle** is an eminence at the cranial part of the glenoid cavity. The tubercle shows a slight medial inclination on which a small tubercle, the **coracoid process,** can be distinguished. The coracobrachialis arises from the coracoid process, while the biceps brachii arises from the supraglenoid tubercle.

HUMERUS

The humerus (Fig. 5) is located in the arm or brachium. This bone enters into the formation of both the shoulder and elbow. The shoulder is formed by the articulation of the scapula and humerus, and the elbow by the articulation of radius and ulna with each other and with the humerus. The proximal extremity of the humerus includes the head, neck and the greater and lesser tubercles. The distal extremity, the condyle, includes the trochlea, capitulum, and the radial and olecranon fossae, which communicate proximal to the trochlea through the supratrochlear foramen. The medial and lateral epicondyles are situated on the sides of the condyle. The **body** of the humerus lies between the two extremities.

The **head** of the humerus is the part that articulates with the scapula. It presents more than twice the area of the glenoid cavity of the scapula and is elongated sagittally. Although the shoulder is a typical ball-and-socket joint, it normally undergoes only flexion and extension. The **intertubercular groove** begins at the cranial end of the articular area. It lodges the tendon of origin of the biceps brachii and is deflected toward the median plane by the **greater tubercle** which forms the craniolateral part of the proximal extremity. The greater tubercle is convex at its summit and, in most breeds, higher than the head. It

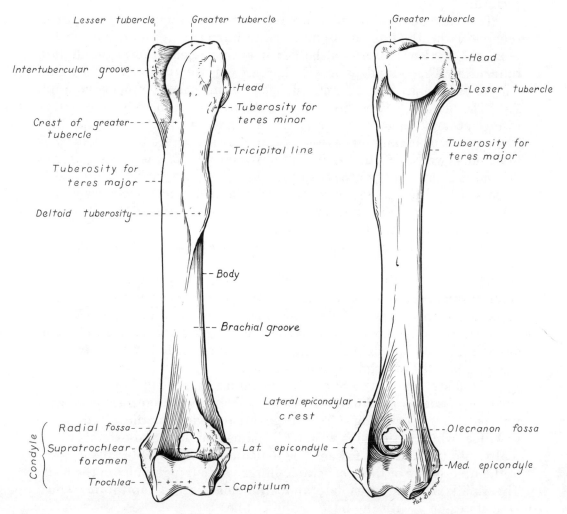

Figure 5. Left humerus, cranial view left and caudal view right.

receives the insertions of the supraspinatus and the infraspinatus and part of the deep pectoral. Between the head of the humerus and the greater tubercle are several foramina for the transmission of vessels. The infraspinatus is inserted on the smooth facet of the lateral side of the greater tubercle. The **lesser tubercle** lies on the medial side of the proximal extremity of the humerus, caudomedial to the intertubercular groove. It is not as high or as large as the greater tubercle. To its proximal border attaches the subscapularis. The neck of the humerus is not distinct except caudally. It is the line along which the head and parts of the tubercles have fused with the body.

The **cranial border** of the humerus is distinct in the middle third of the body, where it furnishes attachment for the brachiocephalicus and part of the pectorals. Distally it fades but may be considered to continue to the medial lip of the trochlea. On the proximal third of the cranial border are two ridges. They continue to the craniomedial and caudolateral extremities of the greater tubercle. The ridge which extends proximally in a craniomedial direction is the **crest of the greater tubercle** and is also the cranial border of the bone. This forms part of the area of insertion of the pectorals.

The ridge extending to the caudolateral extremity of the greater tubercle is on the **lateral surface** of the humerus. At its origin on the proximal third of the cranial border it is thickened to form the **deltoid tuberosity.** The deltoideus inserts here. From this tuberosity to the caudal part of the greater tubercle the ridge forms the prominent **tricipital line.** The lateral head of the triceps arises from this line. The teres minor inserts on and adjacent to the proximal extremity of the tricipital line. The smooth **brachialis groove** is on the lateral surface of the body. The brachialis, which originates in the proximal part of the groove, spirals around the bone in the groove so that distally it lies on the craniolateral surface. Distal to this groove is the thick **lateral epicondylar crest.** The extensor carpi radialis and part of the anconeus attach here. The crest extends distally to the lateral epicondyle.

The **caudal border** is smooth and rounded transversely and ends in the deep olecranon fossa.

The **crest of the lesser tubercle** crosses the proximal end of the **medial surface** and ends distally at the **major teres tuberosity.** The teres major and latissimus dorsi are inserted on this tuberosity. Behind and above this the medial head of the triceps arises and the coracobrachialis is inserted. Approximately the middle third of the medial surface is free of muscular attachment and is smooth.

The distal end of the humerus, including its articular areas and the adjacent fossae, is the **humeral condyle.** The articular surface is divided unevenly by a low ridge. The large area medial to the ridge is the **trochlea,** which articulates with both radius and ulna and extends proximally into the adjacent fossae. The articulation with the trochlear notch of the ulna is one of the most stable hinge joints (ginglymus) in the body. The small articular area lateral to the ridge is the **capitulum,** which articulates only with the head of the radius.

The **lateral epicondyle** is smaller than the medial one and occupies the enlarged distolateral end of the humerus proximal to the capitulum. It gives origin to the common digital extensor, lateral digital extensor ulnaris lateralis and supinator. The lateral ligament of the elbow also attaches here. The lateral epicondyloid crest extends proximally from this epicondyle.

The **medial epicondyle** is the enlarged distomedial end of the humerus proximal to the trochlea. Its caudal projection deepens the olecranon fossa. The anconeus arises from this projection. The elevated portion of the medial epicondyle serves as origin for flexor carpi radialis, flexor carpi ulnaris, pronator teres and the superficial and deep digital flexor muscles. The medial ligament of the elbow also attaches here.

The **olecranon fossa** is a deep excavation of the caudal part of the humeral condyle. It receives the anconeal process of the ulna during extension of the elbow. On the cranial surface of the humeral condyle is the **radial fossa,** which communicates with the olecranon fossa by an opening, the **supratrochlear foramen.** No soft structures pass through this foramen.

RADIUS

The radius and ulna are the bones of the antebrachium or forearm. It is important to know that they cross each other obliquely so that the proximal end of the ulna is medial and the distal end lateral to the radius. The radius (Fig. 6), the shorter of the two bones of the forearm, articulates proximally with the humerus and distally with the carpus. It also articulates with the ulna, proximally by its caudal surface and distally near its lateral border.

The proximal extremity consists of head, neck and tuberosity. The **head** of the radius, like the whole bone, is compressed from before backward. It forms proximally an oval, depressed articular surface, the **fovea capitis,** which articulates with the capitulum of the humerus. The small **radial tuberosity** lies distal to the neck on the medial border of the bone. Parts of the brachialis and biceps brachii are inserted on this tubercle.

The **body** of the radius is compressed so that it possesses cranial and caudal surfaces and medial and lateral borders. It is slightly convex cranially. At the carpal end the shaft blends without sharp demarcation with the enlarged distal extremity. The caudal surface of the radius is roughened and slightly concave. It broadens distally and becomes the expanded caudal surface of the distal extremity. The cranial surface of the radius, convex transversely, is relatively smooth throughout.

The distal extremity of the radius is the **trochlea** Its carpal articular surface is concave. On the lateral surface of the distal extremity is the **ulnar notch,** a slightly concave area with a facet for articulation with the ulna. This notch does not face directly laterally, but partly caudally. The medial surface of the distal extremity ends in a rounded projection, the **styloid process.** The medial ligament of the carpus attaches proximal to the styloid process. The cranial surface of the distal extremity presents

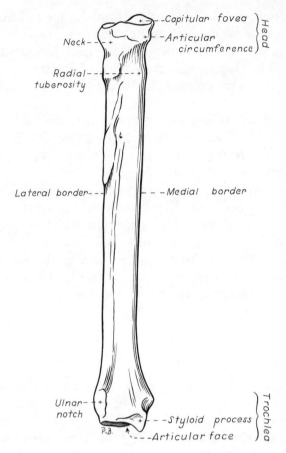

Figure 6. Radius, caudal view.

three distinct grooves. The most medial groove, small, short and oblique, contains the tendon of the abductor pollicis longus. The middle and longest groove, extending proximally on the shaft of the radius, is for the extensor carpi radialis. The most lateral of the grooves on this surface is wide and of variable distinctness. It contains the tendon of the common digital extensor.

ULNA

The ulna (Fig. 7) is located in the caudal part of the forearm. It exceeds the radius in length and is irregular in shape and generally tapers from its proximal to its distal end. Proximally the ulna articulates with the trochlea of the humerus by the trochlear notch and with the articular circumference of the radius by the radial notch. This forms the elbow. Distally the ulna articulates with the radius and with the ulnar and accessory carpal bones.

The proximal extremity is the **olecranon.** It serves as a lever arm for the extensor muscles of the elbow. It is four-sided, laterally compressed and medially inclined. Its proximal end is grooved cranially and enlarged and rounded caudally. The triceps brachii, anconeus and tensor fasciae antebrachii attach to the caudal part of the olecranon. The ulnar portions of the flexor carpi ulnaris and deep digital flexor arise from its medial surface.

The **trochlear notch** is a smooth, vertical, half-moon-shaped concavity facing cranially. The whole trochlear notch articulates with the trochlea of the humerus. At its proximal end, a sharp-edged, slightly hooked **anconeal process** fits into the olecranon fossa when the elbow is extended. At the distal end of the notch is the **coronoid process.** It articulates with the humerus and radius.

The **body** of the ulna is three-sided in its middle third; proximal to this the bone is compressed laterally, while the distal third gradually loses its borders, becomes irregular and is continued by the pointed distal extremity. The **ulnar tuberosity** is a small, elongated eminence on the medial surface of the bone at its proximal end, just distal to the medial coronoid process. The biceps brachii and the brachialis insert on this eminence. The **interosseous border** is distinct, rough and irregular,

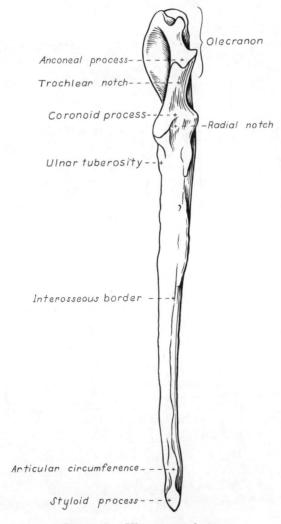

Figure 7. Ulna, cranial view.

especially at the junction of the proximal and middle thirds of the bone, where a large, expansive but low eminence is found. This eminence indicates the place of articulation with the radius by means of a heavy ligament. Frequently a vascular groove medial to the crest marks the position of the caudal interosseous artery in life. This groove is most conspicuous in the middle third of the ulna. The body shows a distinct caudal concavity.

The distal extremity of the ulna is the **styloid process.** A part of this articulates with the ulnar and accessory carpal bones.

CARPAL BONES

The term **carpus** (Figs. 8, 9) is used to designate that part of the extremity between the forearm and metacarpus that includes all the soft

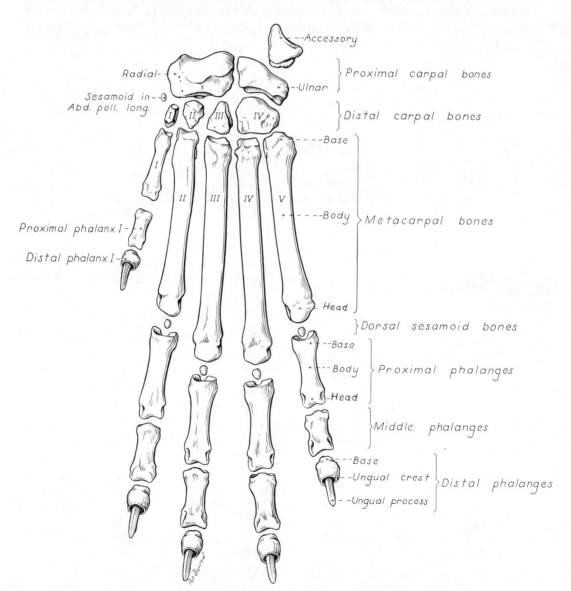

Figure 8. Bones of forepaw, dorsal view.

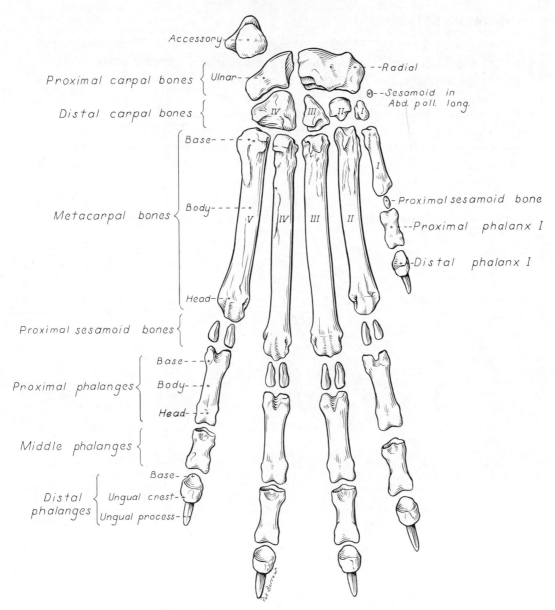

Figure 9. Bones of forepaw, palmar view.

structures as well as the bones. The carpus includes seven small, irregular bones arranged into two rows. The proximal row contains three bones. The largest of these, the **radial carpal,** is on the medial side and articulates proximally with the radius. It represents the fused radial, central and intermediate carpal bones of the fetus. The **ulnar carpal** is the lateral member of the proximal row. The **accessory carpal,** the palmar member, is a short rod of bone which articulates with the styloid process of the ulna and the ulnar carpal bone and serves as a lever arm for some of the flexor muscles of the carpus. The distal row consists of four bones numbered from the medial to the lateral side. From the smallest on the medial side, these are the first, second, third and fourth carpal bones.

METACARPAL BONES

The **metacarpus** (Figs. 8, 9) contains five bones. The metacarpal bones are long bones in miniature, possessing a slender **body** or shaft and large extremities. The proximal extremity is the **base,** the distal one the **head.** The metacarpals, like the carpals and digits, are numbered from medial to lateral. Proximally all articulate principally with the corresponding carpal bones, except the fifth, which articulates with the fourth carpal. Distally all articulate with the corresponding proximal phalanges. The interossei largely fill the intermetacarpal spaces caudal to the metacarpal bones.

The first metacarpal bone is atypical. It is a vestigial structure but, unlike the first metatarsal bone in the hind paw, is constantly present.

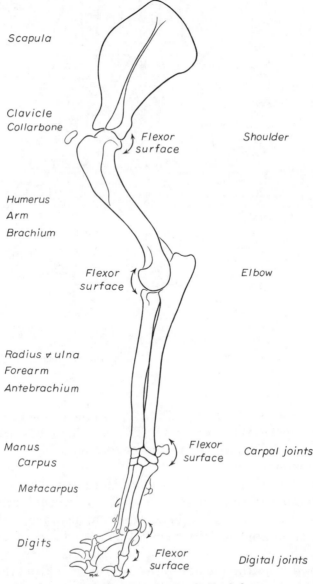

Figure 10. Parts of forelimb skeleton.

PHALANGES

In the forepaw there are three phalanges (Figs. 8, 9) for each of the four main digits; the first digit or pollex, a dewclaw, has two phalanges. Each phalanx has a proximal **base** and distal **head.**

On the distal or third phalanx, a lip projects proximally from the palmar border of the base, upon which the deep digital flexor is inserted. A thin shelf of bone, the **ungual crest,** overlaps the claw distally and forms a band of bone around the root of the claw. The rounded part of the base dorsally is the **extensor process** on which the common digital extensor muscle is inserted. The **ungual process** is a curved conical extension of the distal phalanx into the claw.

Two proximal sesamoid bones are located in the interosseous tendons on the palmar surface of each metacarpophalangeal joint (digits two to five). Four small dorsal **sesamoid bones** (none for the first digit) are imbedded in the extensor tendons as they pass over the metacarpophalangeal joints (Figs. 8, 9).

MUSCLES OF THE THORACIC LIMB

SUPERFICIAL STRUCTURES OF THE THORAX

General study of the ventral surface of the trunk should be made before the dissection of the thoracic region is begun. Find the **umbilicus.** This is represented by a scar which may be either flat or slightly raised and is located on the midventral line, from one-third to one-fourth of the distance from the xiphoid cartilage to the scrotum or vulva. In most dogs it is hidden by hair. The umbilicus is irregularly oval and may be from a few millimeters to a centimeter in length. The umbilicus serves as a landmark in abdominal surgery. Notice that the hair over a large area around the umbilicus slants toward it, thus forming a **vortex.**

Pick up a fold of skin or **common integument.** Notice that it is thickest in the neck region, thinner over the sternum and thinnest on the ventral surface of the abdomen; also notice that the skin of the dorsum of the neck and thorax is loosely attached.

The mammae vary in number from 8 to 12 but 10 is the average number. They are situated in two rows, usually opposite each other. The number is most often reduced in the smaller breeds.

When 10 glands are present the cranial four are the **thoracic mammae,** the following four, the **abdominal mammae** and the caudal two, the **inguinal mammae.** When the abdominal and inguinal mammae are maximally developed the glandular tissue in each row appears to form a continuous mass. It should be observed that the mammae lie in the areolar connective tissue. The cranial pair of thoracic mammae are smaller than the other pair. Each mamma has a papilla which is partly hairless and contains about 12 openings, but these vary and are difficult to see if the animal is not lactating.

The costal cartilages of the tenth, eleventh and twelfth ribs unite with each other to form the **costal arch.** Palpate this arch and the caudal border and free end of the last or thirteenth rib. This rib does not attach to the costal arch and is therefore a floating rib.

Make a midventral incision *through the skin only* from the cranial end of the neck to the umbilicus (Fig. 11). From the umbilicus extend a transverse incision to the middorsal line on the left side. From a point on the midventral incision directly opposite the arm extend a transverse incision to the left elbow. Make a circular incision through the skin completely around the elbow. Extend a third transverse incision from the cranial end of the midventral incision to the middorsal line on the left side. This should pass just caudal to the ear. Carefully reflect the skin of the thorax and neck to the middorsal line. The skin will be intimately fused with the thin underlying **cutaneus trunci** muscle. The latter should be left on the specimen as far as is possible.

The subcutaneous tissue which now confronts the dissector is composed of areolar tissue, a loose, irregularly arranged connective tissue, and superficial fascia, a denser, regularly arranged connective tissue that is closely applied to the superficial muscles. The areolar tissue contains fat and may be distended with embalming fluid. Subcutaneous injections are made into this tissue. When the fascia, fat, vessels and nerves are removed from a muscle, the muscle is said to be cleaned.

In many instances during the study of the muscles a description of a specific muscle will be given before the instructions for dissection. At no time should muscles be removed or even transected without instructions. In each instance clean the exposed surface of the muscle being described, isolate its borders, and verify its origin and insertion. If the muscle is to be transected, free it from underlying structures first.

The **cutaneus trunci** (see Fig. 15) is a thin, muscular sheet which covers most of the dorsal, lateral and ventral walls of the thorax and abdomen. It is more closely applied to the skin than to underlying structures. Like all cutaneous muscles, it is developed in the superficial fascia of the thorax and abdomen. Behind the shoulder the fibers sweep

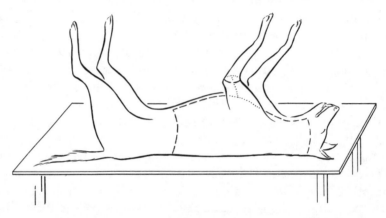

Figure 11. Dissecting position and first skin incisions.

obliquely toward the axilla; farther caudally they are principally longitudinal and arise from the superficial fascia over the rump.

The attachments of the cutaneus trunci are the superficial fascia of the trunk and the skin. The muscle sends a fasciculus to the medial side of the forelimb; caudal and ventral to this the fibers fray out over the deep pectoral muscle. The cutaneus trunci twitches the skin.

Sever the axillary and ventral attachments of the cutaneus trunci and reflect it dorsally. Caution: Beneath the cutaneous trunci is the latissimus dorsi muscle, which should be left in place on the lateral side of the trunk.

Extrinsic Muscles of the Thoracic Limb and Related Structures

The extrinsic muscles of the thoracic limb are those that attach the limb to the axial skeleton; the intrinsic muscles extend between the bones which compose the limb itself.

In the ventral thoracic region are the superficial and deep pectoral muscles which extend between the sternum and the arm. Thoroughly clean these muscles. In thin specimens this will require little dissecting. In pregnant or lactating bitches it will require reflecting the two thoracic mammae caudally, while in fat specimens the forelimb will probably have to be manipulated so that the borders of the muscles are clearly discernible before cleaning. Always clean the extremities of a muscle as well as the middle part. Actually see and feel the attachments. Visualize the muscle's position and action on the skeleton.

1. The two **superficial pectoral muscles** (Figs. 12-14) lie under the skin between the cranial part of the sternum and the humerus. Their caudal border is thin; their cranial border is thick and rounded and forms the caudal border of a triangle at the base of the neck. The smaller **descending pectoral** is superficial to the transverse pectoral, which it obliquely crosses from its origin on the first sternebra to its insertion on the crest of the greater tubercle of the humerus. The **transverse pectoral** arises from the first two or three sternebrae and inserts over a longer distance on the crest of the greater tubercle of the humerus. It is related on its deep surface to the deep pectoral muscle (ascending pectoral). At their insertions these muscles lie between the brachiocephalicus in front and the biceps brachii and humerus behind. Clean both of these superficial pectoral muscles. Transect them 1 cm. from the sternum and reflect them toward the humerus.

As muscle attachments are being cleaned, examine the skeletal parts involved.

ORIGIN: The first two sternebrae and usually a part of the third; the fibrous raphe between fellow muscles.

INSERTION: The whole crest of the greater tubercle of the humerus.

ACTION: To adduct the limb when it is not bearing weight or to prevent the limb from being abducted when bearing weight.

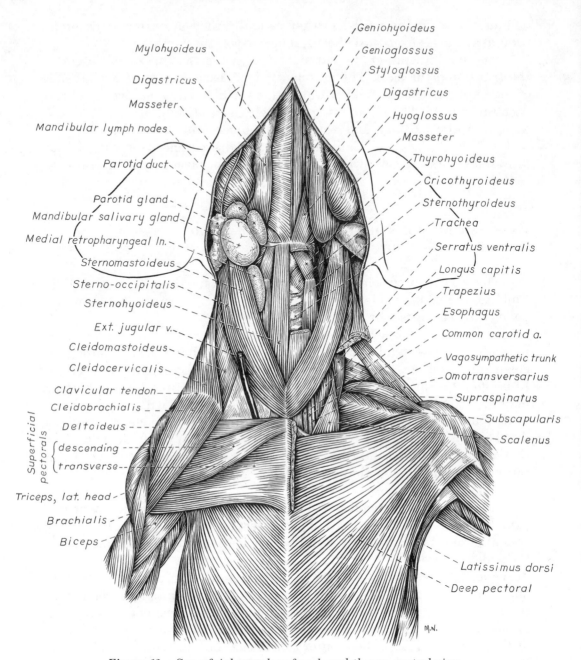

Figure 12. Superficial muscles of neck and thorax, ventral view.

2. The **deep pectoral muscle.** It extends from the sternum to the humerus and is larger and longer than the superficial pectoral muscles. It lies largely under the skin and the thoracic mammae. The papilla of the caudal thoracic mamma usually lies at the caudal border of the muscle. Only the cranial part is covered by the superficial pectoral muscles. An abdominal slip of this muscle is often present on the caudolateral border. Transect the deep pectoral muscle 2 cm. from and parallel to the sternum and clean the distal part to its insertion.

ORIGIN: The ventral part of the sternum and the fibrous raphe between fellow muscles; the deep abdominal fascia in the region of the xiphoid cartilage (the caudal end of the sternum).

INSERTION: The major portion partly muscular, partly tendinous on the lesser tubercle of the humerus; an aponeurosis to the greater tubercle and its crest; the caudal part to the medial brachial fascia.

ACTION: To pull the trunk cranially when the limb is advanced and fixed; extend the shoulder; draw the limb caudally when it is not supporting weight.

The **superficial fascia of the neck** is continued on the head as the superficial fascia of the various regions of the head. Caudally, it becomes continuous with the superficial brachial and pectoral fasciae. Some of the fascia is also continued into the axillary space. The cutaneous muscles of the neck are completely enveloped by this fascia. Notice that the external jugular vein is completely wrapped by this fascia. Save this vein for future orientation.

3. The **platysma** (Fig. 170) is the best developed of the cutaneous muscles of the neck and head. Notice its fibers sweeping cranioventrally over the dorsal part of the neck toward the lateral surface of the face. Note its relationship to the external jugular vein. Transect this muscle across the neck in the same plane as the skin was transected and reflect it to the middorsal line. The muscle may have been reflected with the skin. There are variably developed sphincter colli muscles, also located in the superficial fascia, which will not be dissected.

Reflect the skin cranially to see the flat, loosely lobulated **parotid salivary gland** (Figs. 12, 171) at the base of the ear. The gland and its duct (which crosses the masseter muscle) will be dissected later. Ventral to the parotid is the ovoid, encapsulated **mandibular salivary gland** which lies between the two large veins forming the external jugular. Ventrocranial to the mandibular salivary gland are from two to four **mandibular lymph nodes.** Transect the external jugular vein caudal to the mandibular gland and carefully loosen en masse the caudal parts of the parotid and mandibular glands, the mandibular lymph nodes and the binding fascia. Reflect them cranially to expose the attachments of parts of the following two muscles.

4. The **brachiocephalicus** (Figs. 12-15) of the dog is a compound muscle developmentally although it appears as one muscle which extends from the arm to the head and neck. One end attaches on the distal third of the humerus, where it lies between the biceps brachii medially and the brachialis laterally. Proximally on the humerus it partly covers the pectoral muscles at their insertions and lies craniomedial to the deltoid muscle. It crosses the cranial surface of the shoulder, divides into two parts and obliquely traverses the neck. At the shoulder a faint line crosses the muscle. This is the edge of a fibrous plate, the **clavicular intersection,** on the deep surface of which the vestigial clavicle (collar bone) is connected. Although the clavicle has lost its functional significance in the dog, it is

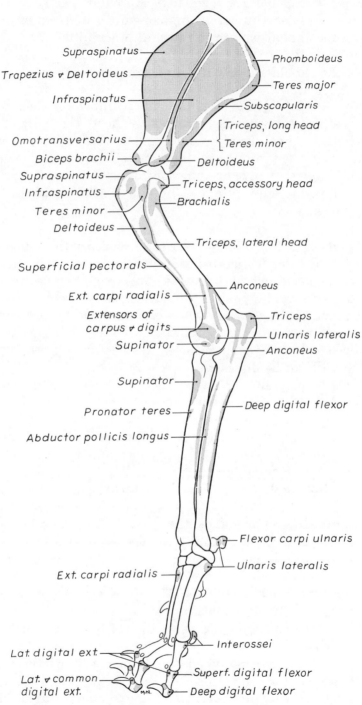

Figure 13. Forelimb skeleton, lateral view of muscle attachments.

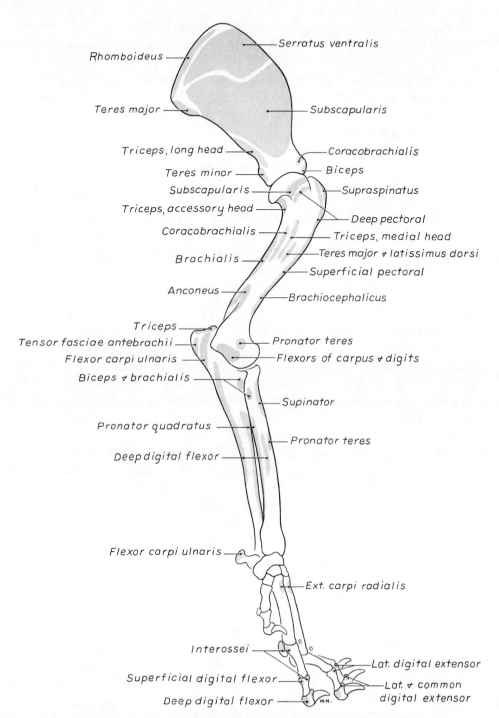

Figure 14. *Forelimb skeleton, medial view of muscle attachments.*

still considered the origin of the components of the brachiocephalicus muscle. Thus the muscle distal to the clavicular tendon that attaches to the arm is the **cleidobrachialis,** the muscle that extends from the clavicular tendon to the dorsum of the neck is the **cleidocervicalis** and beneath this is the **cleidomastoideus,** which attaches to the skull. The cleidocervicalis is bounded caudally by the trapezius and cranially by the sternocephalicus.

Transect the cleidocervicalis in order to expose the full extent of the cleidomastoideus. Note that the cleidomastoideus runs toward the head deep to the sternocephalicus. Transect the cleidomastoid muscle and search for the clavicle by inserting your finger on the medial side of the clavicular intersection.

ATTACHMENTS: All attachments are movable but the clavicle or clavicular intersection is considered as the origin for purposes of naming the muscles. The cleidobrachialis attaches to the distal end of the cranial border of the humerus. There is also a significant fascial tie into the axilla. The cleidocervicalis attaches to the cranial half of the middorsal fibrous raphe and sometimes to the nuchal crest of the occipital bone. The cleidomastoideus attaches to the mastoid part of the temporal bone with the sternomastoideus muscle.

ACTION: To advance the limb; draw the neck and head to the side.

5. The **sternocephalicus** (Figs. 12, 15) arises on the sternum and inserts on the head. At the cranial end of the sternum the muscle is thick, rounded and closely united with its fellow of the opposite side. Even after the main parts of the paired muscle diverge, there is considerable crossing of fibers between the two on the ventral surface of the neck. The dorsal border of the sternocephalicus is adjacent to the ventral border of the brachiocephalicus. The external jugular vein crosses its lateral surface obliquely. Notice that the cranial part of the muscle divides into two parts and that the ventral part is closely related to the cleidomastoid division of the brachiocephalicus. The ventral part of the sternocephalicus, the **sternomastoideus,** is similar to the cleidomastoideus in shape and insertion. It represents the chief continuation of the sternocephalicus to the head. The thin but wide dorsal part of the sternocephalicus is the **sternooccipitalis.**

ORIGIN: The first sternebra or manubrium.

INSERTION: The mastoid part of the temporal bone and the nuchal crest of the occipital bone.

ACTION: To draw the neck and head to the side.

Transect the left sternocephalicus close to the manubrium and reflect it. The sternohyoid and sternothyroid muscles are both covered by the deep fascia of the neck and at their origin lie dorsal to the sternocephalicus.

6. The **sternohyoideus** (Fig. 12) lies on the trachea. A midventral groove indicates the separation of right and left muscles.

ORIGIN: The first sternebra and the first costal cartilage.

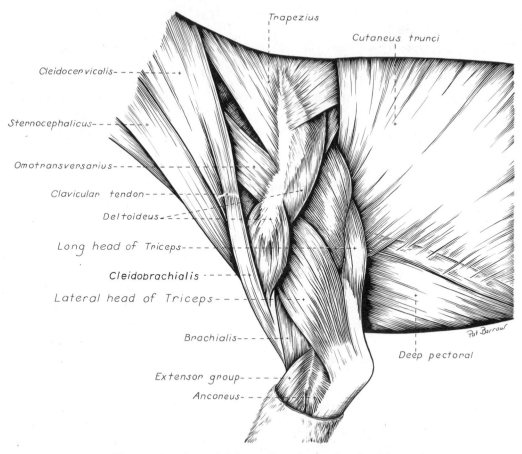

Figure 15. Superficial muscles of scapula, shoulder and arm.

INSERTION: The basihyoid bone.

ACTION: To pull the tongue and larynx caudally.

7. The **sternothyroideus** (Fig. 12) is covered at its origin by the sternohyoideus. The sternothyroideus inserts on the lateral surface of the thyroid cartilage. The left muscle is bounded dorsally by the esophagus and medially by the trachea. Notice that a tendinous intersection runs across the muscle 3 or 4 cm. cranial to its origin. It is at this level that the sternohyoideus separates from the sternothyroideus.

ORIGIN: The first costal cartilage.

INSERTION: The caudolateral surface of the thyroid cartilage.

ACTION: Same as the sternohyoideus—to draw the larynx and tongue caudally.

8. The **omotransversarius** (Figs. 12, 13, 15) is in a deeper plane than the brachiocephalicus. It is straplike and extends from the distal end of the spine of the scapula to the atlas. It is related to the deep cervical fascia medially. Its caudal part is subcutaneous, but farther forward it is covered by the cervical part of the brachiocephalicus. Transect the omotransversarius through its middle and reflect each half toward its attach-

ment. This will expose the **superficial cervical lymph nodes** located cranial to the scapula.

ATTACHMENTS: The distal end of the spine of the scapula; and cranially, the wing of the atlas.

ACTION: To advance the limb or flex the neck laterally.

The **deep fascia of the neck** is a strong wrapping which extends under the sternocephalicus, cleidomastoideus, omotransversarius and cleidocervicalis muscles. It covers the sternohyoideus and sternothyroideus ventrally and surrounds the trachea, thyroid gland, larynx and esophagus. The deep fascia that covers the common carotid artery, vagosympathetic nerve trunk and internal jugular vein is the **carotid sheath.** Locate these structures in the carotid sheath between the omotransversarius dorsally and the sternothyroideus ventrally. The deep fascia of the neck continues dorsally and laterally to invest the deep cervical muscles.

9. The **trapezius** (Figs. 13, 15) is thin and triangular. It is divided into cervical and thoracic parts, separated by an aponeurosis. The muscle as a whole extends from the median raphe of the neck and the supraspinous ligament to the spine of the scapula. The cervical part is overlapped by the cleidocervicalis, while the thoracic part overlaps the latissimus dorsi. Make an arching cut through the muscle, beginning at the middle of the cranial border over the dorsal border of the scapula, through the aponeurotic area, to the middle of the caudal border. Reflect the muscle to its attachments.

ORIGIN: The median raphe of the neck and the supraspinous ligament from the level of the third cervical vertebra to the level of the ninth thoracic vertebra.

INSERTION: The spine of the scapula.

ACTION: To elevate and abduct the forelimb.

10. The **rhomboideus** (Figs. 13, 14, 16) lies beneath the trapezius and holds the dorsal border of the scapula close to the body. It has capital, cervical and thoracic parts. The narrow **rhomboideus capitis** attaches the cranial dorsal border of the scapula to the nuchal crest of the occipital bone. The **rhomboideus cervicis** runs from the median raphe of the neck to the dorsal border of the scapula. The **rhomboideus thoracis** is short and thick. The cervical and thoracic parts of the rhomboideus are contiguous on the dorsal border of the scapula. Transect the entire muscle a few cm. from the scapula.

ORIGIN: The nuchal crest of the occipital bone; the median fibrous raphe of the neck; the spinous processes of the first seven thoracic vertebrae.

INSERTION: The dorsal border and adjacent surfaces of the scapula.

ACTION: To elevate the forelimb.

11. The **latissimus dorsi** (Figs. 14, 16) is large and roughly triangular. It lies caudal to the scapula, where it covers most of the dorsal and some of the lateral thoracic wall. Clean its ventrocaudal border. Directly

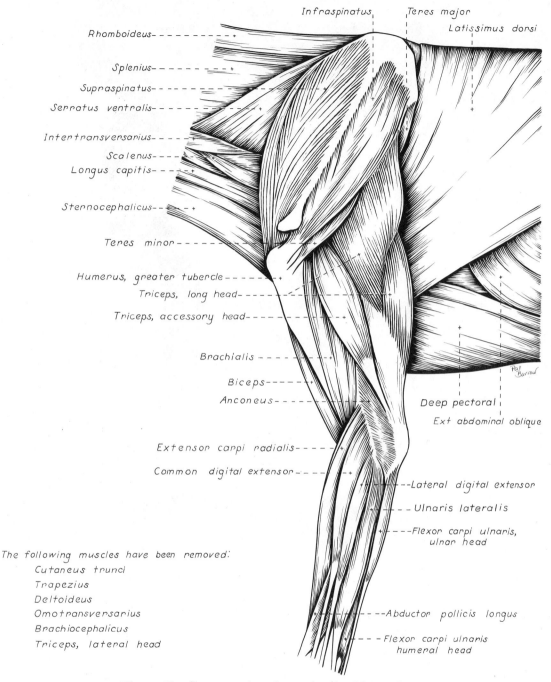

Figure 16. Deep muscles of scapula, shoulder and arm.

caudal to the forelimb, transect the latissimus dorsi at a right angle to its fibers.

ORIGIN: The thoracolumbar fascia from the spinous processes of the lumbar and the last seven or eight thoracic vertebrae; a muscular attachment to the last two or three ribs.

INSERTION: The major teres tuberosity of the humerus.

ACTION: To draw the free limb caudally as in digging; to flex the shoulder.

12. The **serratus ventralis** (Fig. 14, 16) is a powerful, fan-shaped muscle which acts as a sling to support the body between the limbs. Abduct the forelimb. This will require severing the axillary artery and vein, the brachial plexus of nerves and the axillary fascia. As the forelimb is progressively abducted the attachment of the serratus ventralis will detach from the serrated face of the scapula. Since the forelimb is removed at this stage of the dissection, the detachment of the serratus ventralis may be completed. It is the only extrinsic muscle of the forelimb that has not been transected.

ORIGIN: The transverse processes of the last five cervical vertebrae and the first seven or eight ribs ventral to their middle.

INSERTION: The dorsomedial third of the scapula (serrated face).

ACTION: To support the trunk and depress the scapula.

INTRINSIC MUSCLES OF THE THORACIC LIMB

LATERAL MUSCLES OF THE SCAPULA AND SHOULDER

1. The **deltoideus** (Figs. 13, 15) is composed of two portions which fuse and act in common across the shoulder. The proximal portion

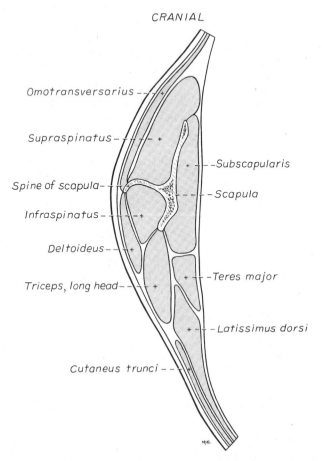

Figure 17. Cross-section through middle of left scapula.

arises as a wide aponeurosis from the length of the scapular spine and covers the infraspinatus. The latter muscle can be seen through this aponeurosis and should not be confused with the deltoideus. Observe the distal portion of the deltoideus, which arises from the acromion and has a fusiform shape. Both portions of the muscle fuse before they insert on the humerus. Transect the combined muscle 2 cm. distal to the acromion and reflect the stumps. Work under the aponeurosis of origin to verify its attachment to the spine of the scapula.

ORIGIN: The spine and acromial process of the scapula.

INSERTION: The deltoid tuberosity.

ACTION: To flex the shoulder.

2. The **infraspinatus** (Figs. 13, 16, 17) is fusiform and lies principally in the infraspinous fossa. Transect the infraspinatus halfway between its extremities and free the distal half from the scapula by scraping the fibers away from the spine and fossa with the handle of the scalpel. Reflect the distal half to its insertion on the side of the greater tubercle. This will expose a bursa between the tendon of insertion and the greater tubercle of the humerus. A bursa is a closed sac containing synovial fluid which reduces friction.

ORIGIN: The infraspinous fossa.

INSERTION: A small, circumscribed area distal to the greater tubercle of the humerus.

ACTION: To extend, flex or abduct the joint, depending on the degree of extension or position of this joint when the muscle contracts. It also acts as a collateral ligament.

3. The **teres minor** (Figs. 13, 16), a small, wedge-shaped muscle is now exposed caudal to the shoulder. It is covered superficially by the deltoideus, caudally by the triceps and cranially by the infraspinatus.

ORIGIN: The infraglenoid tubercle and distal third of the caudal border of the scapula.

INSERTION: The minor teres tuberosity of the humerus.

ACTION: To flex the shoulder.

4. The **supraspinatus** (Figs. 13, 14, 16, 17), wider and larger than the infraspinatus, is largely covered by the cervical part of the trapezius and the omotransversarius. It lies in the supraspinous fossa and extends over the cranial border of the scapula so that a part of the muscle is closely united with the subscapularis. Clean and observe the insertion on the greater tubercle of the humerus.

ORIGIN: The supraspinous fossa.

INSERTION: The greater tubercle of the humerus, by a strong tendon.

ACTION: To extend the shoulder.

MEDIAL MUSCLES OF THE SCAPULA AND SHOULDER

1. The **subscapularis** (Figs. 13, 14, 17, 19) occupies the entire subscapular fossa, the boundaries of which it overlaps slightly. The supraspina-

tus is closely associated with it cranially, while the teres major has a similar relation caudally. Clean the insertion but do not transect the muscle.

ORIGIN: The subscapular fossa.

INSERTION: The lesser tubercle of the humerus.

ACTION: To adduct and extend the shoulder.

The subscapularis and infraspinatus function actively as the medial and lateral ligaments of the shoulder. To these is added the supraspinatus.

2. The **teres major** (Figs. 13, 14, 19), directly caudal to the subscapularis, belies its descriptive name since it is not round but has three surfaces. From its proximal end, which arises from the subscapularis and the proximal caudal border of the scapula, fibers extend distally to attach to the tendon of insertion of the latissimus dorsi. Work between the distal half of the muscle and the subscapularis. Observe the close relationship between the teres major and the latissimus dorsi. Transect the teres major and reflect the combined insertion to expose the belly of the coracobrachialis muscle.

ORIGIN: The caudal angle and adjacent caudal border of the scapula; the caudal surface of the subscapularis.

INSERTION: The major teres tuberosity of the humerus.

ACTION: To flex the shoulder.

3. The **coracobrachialis** (Figs. 14, 19) crosses the medial surface of the shoulder obliquely. It is a small, spindle-shaped muscle which arises from the coracoid process of the scapula by a relatively long tendon which courses caudodistally across the lesser tubercle. Here it crosses the tendon of insertion of the subscapularis. It is provided with a synovial sheath as it crosses the lesser tubercle of the humerus. The conjoined tendon of the teres major and latissimus dorsi crosses its insertion. Notice that the coracobrachialis tendon courses cranial to the center of the shoulder. This accounts for its action as an extensor muscle of the shoulder. Free the coracobrachialis and isolate the tendon of origin by cutting into its synovial sheath.

ORIGIN: The coracoid process of the scapula.

INSERTION: The crest of the lesser tubercle of the humerus proximal to the major teres tuberosity.

ACTION: To adduct and extend the shoulder.

CAUDAL MUSCLES OF THE ARM (BRACHIUM)

This group comprises a large, muscular mass that almost completely fills the space between the caudal border of the scapula and the olecranon. It consists of three muscles: the triceps brachii, the tensor fasciae antebrachii and the anconeus. By far the largest of these muscles is the triceps. All of the caudal muscles of the arm are extensors of the elbow.

1. The **tensor fasciae antebrachii** (Figs. 14, 19) is a thin strap which

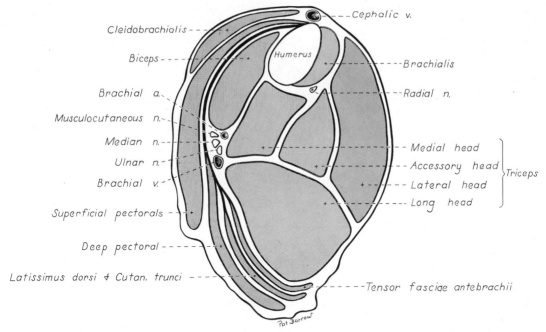

Figure 18. *Cross-section through middle of right arm.*

extends from the latissimus dorsi to the medial fascia of the forearm and the olecranon. It lies on the long head of the triceps.

ORIGIN: The fascia covering the lateral side of the latissimus dorsi.

INSERTION: The olecranon.

ACTION: To extend the elbow.

2. The **triceps brachii** in the dog consists of four heads, instead of the usual three, with a common tendon to the olecranon. Only the long head arises from the scapula. The other three arise from the proximal end of the humerus.

The **long head** (Figs. 13–19) completely bridges the humerus since it arises from the caudal border of the scapula and inserts on the olecranon. It appears to have two bellies. A groove may be palpated between the long and lateral heads of the triceps at the level of the deltoideus transection and the insertion of the teres minor. Separate these two heads along this groove. Expose the tendon of the long head and notice the bursa between it and the groove of the olecranon. Notice how the tendons of the other heads blend with that of the long head.

ORIGIN: The caudal border of the scapula.

INSERTION: The olecranon.

ACTION: To extend the elbow and flex the shoulder.

The **lateral head** (Figs. 13, 15, 18) of the triceps lies distal to the long head proximal to the elbow and lateral to the accessory head which it covers. Transect the origin of the lateral head and reflect it to expose the underlying accessory and medial heads.

ORIGIN: The tricipital line of the humerus.

INSERTION: The olecranon.

ACTION: To extend the elbow.

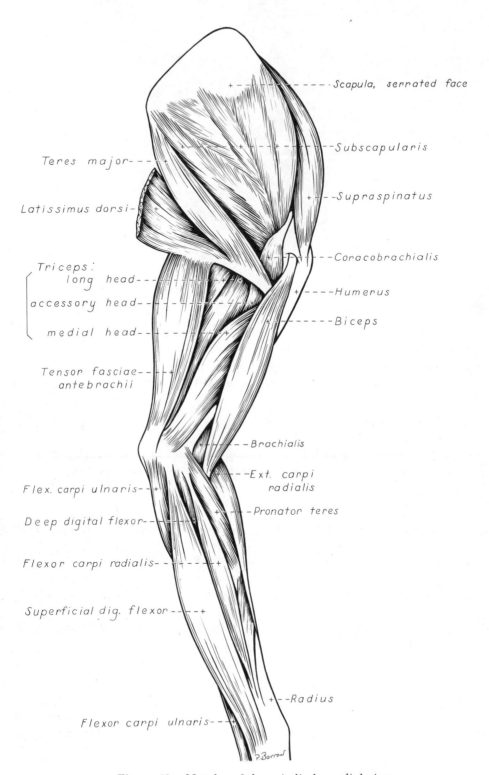

Figure 19. *Muscles of thoracic limb, medial view.*

The **accessory head** (Figs. 13, 14, 16, 18, 19) lies between the lateral and medial heads.

ORIGIN: The neck of the humerus.

INSERTION: The olecranon.

ACTION: To extend the elbow.

The **medial head** (Figs. 14, 18, 19) lies caudally on the humerus. Separate the muscle from the long head caudally and from the accessory head laterally. The long tendon of the accessory head is closely bound to its lateral surface.

ORIGIN: The crest of the lesser tubercle near the major teres tuberosity.

INSERTION: The olecranon.

ACTION: To extend the elbow.

3. The **anconeus** (Figs. 13-16) is a small muscle located almost completely in the olecranon fossa. Reflect the insertion of the lateral head of the triceps to uncover the lateral surface of this muscle. Notice that the most distal fibers lie in a transverse plane. Cut on the lateral side the origin of the anconeus from the lateral epicondyloid crest and epicondyle. Reflect it to expose the elbow joint capsule. Open the joint capsule to expose the elbow.

ORIGIN: The lateral epicondyloid crest and the lateral and medial epicondyles of the humerus.

INSERTION: The lateral surface of the proximal end of the ulna.

ACTION: To extend the elbow.

CRANIAL MUSCLES OF THE ARM

1. The **biceps brachii** (Figs. 12-14, 16, 18, 19) has only one head in domestic animals. It is a long, fusiform muscle which lies on the cranial surface of the humerus. It completely bridges this bone since it arises on the supraglenoid tuberosity and inserts on the proximal ends of the radius and ulna. It is covered superficially by the pectoral muscles. Clean the muscle and transect it through its middle. Reflect the proximal half to its origin. This will require severing the **transverse humeral ligament,** a band of fibrous tissue which joins the greater and lesser tubercles and holds the tendon of origin in the intertubercular groove. An extension of the shoulder joint capsule acts as a synovial sheath for this tendon. Reflect the distal half of the biceps to the proximal end of the radius and ulna, where it meets the brachialis tendon and bifurcates. The tendons of insertion lie on the elbow joint capsule. Delay cleaning these tendons until after the pronator teres muscle has been dissected.

ORIGIN: The supraglenoid tubercle.

INSERTION: The ulnar and radial tuberosities.

ACTION: To flex the elbow and extend the shoulder.

2. The **brachialis** (Figs. 12-16, 18) should be studied from the lateral side. It is a long thin muscle which lies in the brachialis groove of the

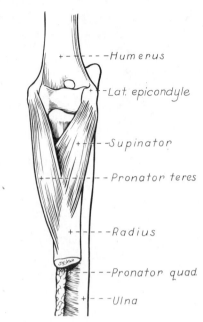

Figure 20. *Rotators of antebrachium.*

humerus. From the proximal third of this groove, the brachialis curves laterally and cranially as it courses distally, crosses the elbow and inserts on the medial side of the proximal end of the ulna. A large part of its lateral surface is covered by the lateral head of the triceps. Distally it runs medial to the origin of the extensor carpi radialis. Its insertion will be dissected later with the biceps insertion.

ORIGIN: The proximal third of the lateral surface of the humerus.

INSERTION: The ulnar and radial tuberosities.

ACTION: To flex the elbow.

Before the rest of the skin is removed from the thoracic limb examine the foot pads. The small pad which protrudes palmar to the carpus is the **carpal pad.** The largest in the paw, the **metacarpal pad,** is on the palmar side of the metacarpal phalangeal joint and is triangular in shape. The **digital pads** are ovoid in outline and flattened. Each is located palmar to the distal interphalangeal joint.

Make a midcaudal incision through the skin from the olecranon through the carpal and metacarpal pads to the interdigital space between digits III and IV. Reflect the skin, dissect it free from the fascia and remove it from the forelimb. The pads are closely attached to the underlying structures but may be dissected free and removed. Be careful not to cut too deeply and sever the underlying small tendons. Work distally on each of the four main digits and completely remove the skin and digital pads.

The subcutaneous tissue distal to the elbow is scanty. As in most other places in the body it connects the skin with the underlying fascia. The principal veins and cutaneous nerves in large part lie on the superficial fascia of the antebrachium. For descriptive purposes the superficial and deep fascia distal to the elbow may be divided into antebrachial, carpal, metacarpal and digital parts.

The **deep antebrachial fascia** forms a single sleeve for the muscles of the forearm. Make an incision through the deep antebrachial fascia from the olecranon to the accessory carpal bone. Carefully reflect the fascia cranially on the forearm. At first the fascia is easily reflected, as it lies on the epimysium of the muscles beneath. Cranially it sends delicate septal leaves between muscles, and on reaching the radius,firmly unites with its periosteum.

CRANIOLATERAL MUSCLES OF THE FOREARM (ANTEBRACHIUM)

The craniolateral antebrachial muscles are, from cranial to caudal, the extensor carpi radialis, the common digital extensor, the lateral digital extensor and the ulnaris lateralis. Most of these muscles arise from the lateral or extensor epicondyle of the humerus.

1. The **extensor carpi radialis** (Figs. 13, 14, 16, 19, 21, 24) is the largest of the craniolateral antebrachial muscles. It lies on the dorsal surface of the radius throughout most of its course and is easily palpated in the live dog. The tendon looks single but is distinctly double throughout its distal third. These closely associated tendons run first under the tendon of the abductor pollicis longus, farther distally in the middle groove of the radius and finally over the carpus. They are held in place by the **extensor retinaculum.** This is a transversely oriented condensation of carpal fascia which aids in holding in grooves all the tendons which cross the dorsum of the carpus. Between bundles of tendons the extensor retinaculum dips down to blend with the fibrous dorsal part of the joint capsule. Define by dissection the proximal and distal margins of the extensor retinaculum, but do not sever it along the tendons.

ORIGIN: The lateral epicondylar crest.

INSERTION: The small tuberosities on the proximal ends and dorsal surfaces of metacarpals II and III.

ACTION: To extend the carpus.

2. The **common digital extensor** (Figs. 13, 21, 24, 25) is shaped like and lies caudal to the extensor carpi radialis on the lateral side. It is smaller than the radial extensor of the carpus and has a multiple tendon of insertion. The four individual tendons which leave the muscle cross the cranial surface of the abductor pollicis longus and then the carpus, where they are held in the lateral groove of the radius by the extensor retinaculum. Distal to the ligament the four tendons diverge and each goes to the third phalanx of one of the four main digits. Dissect the tendon of the common extensor which goes to the fourth digit. Free the tendon as it crosses each of the joints. It is inserted on the extensor process of the third phalanx.

ORIGIN: The lateral epicondyle of the humerus.

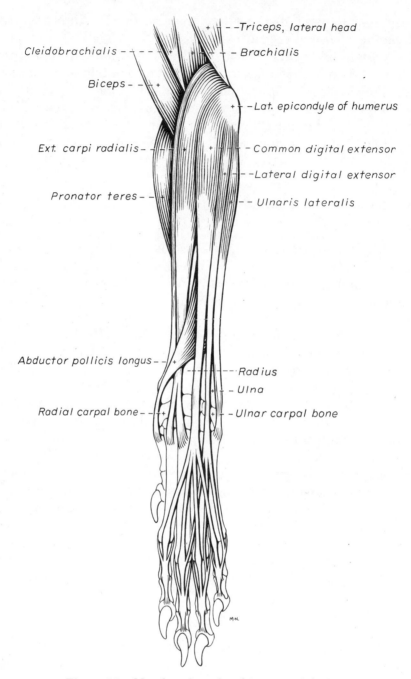

Figure 21. Muscles of antebrachium, cranial view.

INSERTION: The extensor processes of the third phalanges of digits II, III, IV and V.

ACTION: To extend the joints of the four principal digits.

Notice that the distal interphalangeal joint is in a marked degree of overextension. This is brought about by the elastic **dorsal ligament** which lies on each side of the common extensor tendon. This attaches proximally to the sides of the proximal end of the middle phalanx. Distally this

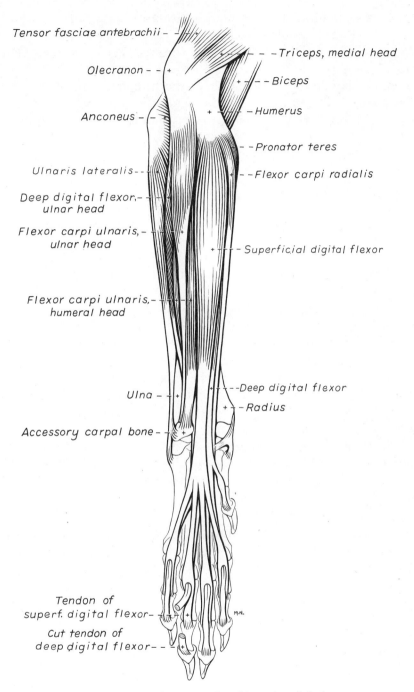

Figure 22. *Muscles of antebrachium, caudal view.*

ligament attaches to the dorsal surface of the ungual crest of the distal phalanx (see Fig. 26). Its elasticity retracts the claw.

3. The **lateral digital extensor** (Figs. 13, 16, 21, 24, 25) is about half the size of the common digital extensor. It lies between the common digital extensor and the ulnaris lateralis. Its tendon begins at the middle of the forearm, passes under the extensor retinaculum in a groove between

the radius and ulna and immediately splits into three branches. The main part of each tendon attaches to the extensor process of the third phalanx of digits III, IV and V in common with the common digital extensor tendon.

ORIGIN: The lateral epicondyle of the humerus.

INSERTION: The proximal ends of all the phalanges of digits III, IV and V, but mainly the extensor processes of the third phalanges of these digits.

ACTION: To extend joints of digits III, IV and V.

4. The **ulnaris lateralis** (Figs. 13, 16, 21, 22, 24, 25) is larger than the lateral digital extensor behind which it lies. It is bounded deeply by the ulna and the large flexor group of muscles which lie behind it. Expose the muscle and notice the two tendons of insertion.

ORIGIN: The lateral epicondyle of the humerus.

INSERTION: The lateral aspect of the proximal end of metacarpal V and the accessory carpal bone.

ACTION: To abduct and flex the carpus.

5. The **supinator** (Figs. 13, 14, 20) is short, broad and flat and obliquely placed across the lateral side of the flexor surface of the elbow joint. It is covered superficially by the extensor carpi radialis and common digital extensors, which should be transected in the middle of their muscle bellies. The supinator lies principally on the proximal fourth of the radius.

ORIGIN: The lateral epicondyle of the humerus.

INSERTION: The cranial surface of the proximal fourth of the radius.

ACTION: To rotate the forearm so that the palmar side of the paw faces medially.

6. The **abductor pollicis longus** (Figs. 13, 16, 21, 24, 25) lies primarily in the groove between the radius and ulna and is triangular. Displace the digital extensors so that the bulk of the muscle is uncovered. Clean the muscle and transect its tendon as it crosses the extensor carpi radialis.

ORIGIN: The lateral border and cranial surface of the body of the ulna; the interosseous membrane.

INSERTION: The proximal end of metacarpals I and II.

ACTION: To abduct the first digit or pollex.

CAUDOMEDIAL MUSCLES OF THE FOREARM

The deep antebrachial fascia has been removed from this group of muscles. When dissecting the individual muscles it will be necessary to clean their tendons of insertion. The muscles in this group include, from the radius caudally, the pronator teres, the flexor carpi radialis, the deep digital flexor, the superficial digital flexor and, on the lateral side, the flexor carpi ulnaris.

1. The **pronator teres** (Figs. 13, 14, 19-22, 24) extends obliquely

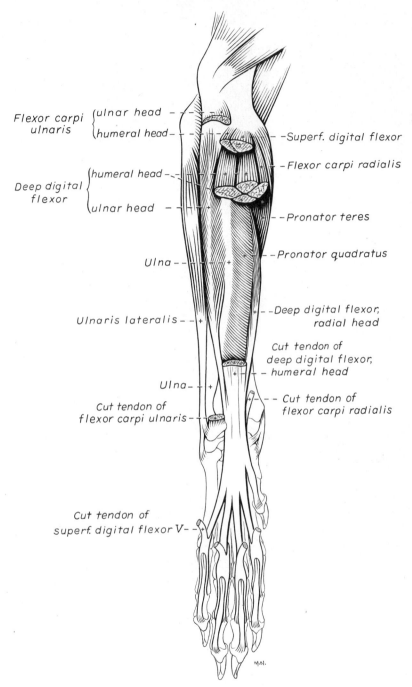

Flexor carpi ulnaris {ulnar head
humeral head}

Deep digital flexor {humeral head
ulnar head}

Ulna

Ulnaris lateralis

Ulna

Cut tendon of flexor carpi ulnaris

Cut tendon of superf. digital flexor V

Superf. digital flexor

Flexor carpi radialis

Pronator teres

Pronator quadratus

Deep digital flexor, radial head

Cut tendon of deep digital flexor, humeral head

Cut tendon of flexor carpi radialis

Figure 23. *Deep muscles of antebrachium, caudal view.*

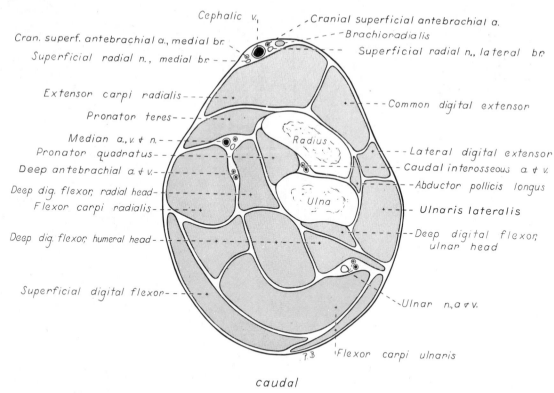

Figure 24. *Cross-section of right forearm between proximal and middle thirds.*

across the medial surface of the elbow. It is round in cross-section at its origin and flat at its insertion. It lies between the extensor carpi radialis cranially and the flexor carpi radialis caudally. Displace adjacent muscles to see its origin and insertion. Transect the muscle and reflect the extremities.

ORIGIN: The medial epicondyle of the humerus.

INSERTION: The medial border of the radius between the proximal and middle thirds.

ACTION: To pronate the forearm and paw; flex the elbow.

Clean the tendons of insertion of the biceps and brachialis muscles, which are now exposed. The tendon of insertion of the biceps splits into two parts. The larger of the two inserts on the ulnar tuberosity and the smaller on the radial tuberosity. The terminal tendon of the brachialis inserts between these two tendons of the biceps on the ulnar tuberosity.

2. The **flexor carpi radialis** (Figs. 14, 19, 22-25) lies between the pronator teres cranially and the superficial digital flexor caudally. It covers the deep digital flexor, part of which can be seen. The flexor carpi radialis has a thick, fusiform belly which, partly imbedded in the deep flexor, extends only to the middle of the radius. Here it gives rise to a flat tendon

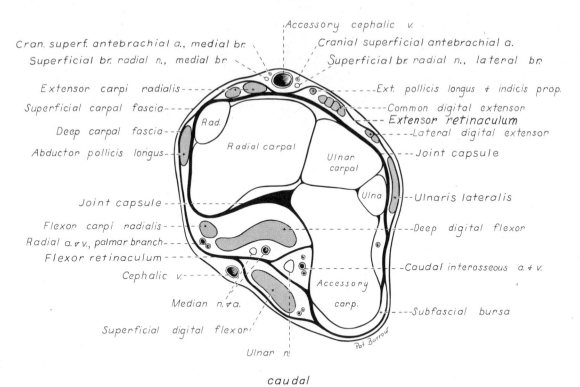

cranial

Accessory cephalic v.

Cran. superf. antebrachial a., medial br.
Cranial superficial antebrachial a.
Superficial br. radial n., medial br.
Superficial br. radial n., lateral br.

Extensor carpi radialis ----
---- Ext. pollicis longus + indicis prop.
Superficial carpal fascia ----
---- Common digital extensor
---- Extensor retinaculum
Deep carpal fascia ----
Rad.
---- Lateral digital extensor
Radial carpal
Abductor pollicis longus ----
---- Joint capsule
Ulnar carpal

Ulna
Joint capsule ----
---- Ulnaris lateralis

Flexor carpi radialis ----
---- Deep digital flexor
Radial a. & v., palmar branch ----
Flexor retinaculum ----
---- Caudal interosseous a. + v.
Cephalic v. ----
Accessory
Median n. & a.
carp.
---- Subfascial bursa
Superficial digital flexor

Ulnar n.

caudal

Figure 25. *Cross-section of right forepaw through accessory carpal bone.*

which is augmented by fibers leaving the medial border of the radius. Clean the tendon to the point where it passes deep to a thick layer of fibrous tissue on the palmar side of the carpus, the **flexor retinaculum** (Fig. 25). Do not cut through this fibrous tissue now. A synovial sheath extends from the distal end of the radius almost to the insertion of the muscle on the second and third metacarpal bones. This will be exposed later.

ORIGIN: The medial epicondyle of the humerus and the medial border of the radius.

INSERTION: The palmar side of the proximal ends of metacarpals II and III.

ACTION: To flex the carpus.

3. The **superficial digital flexor** (Figs. 13, 14, 19, 22-26) lies beneath the skin and antebrachial fascia on the caudomedial side of the forearm. It covers the deep digital flexor and is fleshy almost to the carpus. Its tendon is at first single, then crosses the flexor surface of the carpus medial to the accessory carpal bone where it is covered by the superficial part of the flexor retinaculum, and then divides into four tendons of nearly equal size. These insert on the proximal palmar surfaces of the second phalanges of the four principal digits. At the metacarpophalangeal joint each forms a collar around the deep flexor tendon which passes through it. Clean each of the individual tendons as far as

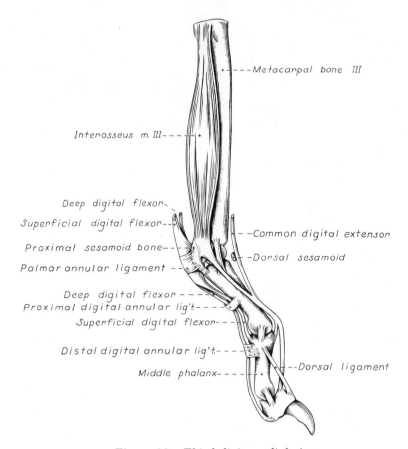

Figure 26. *Third digit, medial view.*

the metacarpophalangeal joints. Transect the muscle at the middle of the antebrachium, and turn the distal part toward the digits. Since all parts of the superficial digital flexor tendon are similar, only that to the third digit will be dissected. The superficial and deep digital flexor tendons are held firmly in place at the metacarpophalangeal joint by the **palmar annular ligament.**

If any of the structures mentioned below are not clearly seen on the third digit, they should be verified on one or more of the other main digits. Observe that the tendon of the superficial digital flexor sheathes the deep digital flexor for a distance of over a centimeter at the metacarpophalangeal joint; that the superficial digital flexor tendon lies on the palmar side of the deep flexor tendon at the proximal end of its encircling sheath but it lies on the dorsal side at the distal end; that the superficial flexor tendon with its sheath and the deep flexor tendon are in a common synovial membrane, the digital synovial sheath.

ORIGIN: The medial epicondyle of the humerus.

INSERTION: The proximal palmar borders of the second phalanges of digits II, III, IV and V.

ACTION: To flex digits II, III, IV and V.

4. The **flexor carpi ulnaris** (Figs. 14, 19, 22–24) consists of two

parts which are distinct throughout their length. The **ulnar head** arises from the palmar border of the proximal end of the ulna. It is thin and wide proximally but narrow distally. It lies between the ulnaris lateralis and superficial digital flexor. The **humeral head** is large and fleshy and lies cranial to the ulnar head, except distally, where its tendon lies caudal to it. Dissect the insertion of this muscle on the accessory carpal bone and clean its origin. A subfascial bursa is present over the tendon of insertion of the humeral head and an intertendinous bursa is found between the two tendons of insertion at the carpus.

ORIGIN: Ulnar head — the palmar border and medial surface of the olecranon; humeral head — the medial epicondyle of the humerus.

INSERTION: The accessory carpal bone.

ACTION: To flex the carpus.

5. The **deep digital flexor** (Figs. 13, 14, 19, 22-25, 27) has three heads of origin which arise from the humerus, radius and ulna. Their bellies, along with the pronator quadratus, lie on the caudal surfaces of the radius and ulna. Transect both muscle bellies of the flexor carpi ulnaris in the middle of the antebrachium. Reflect the stumps to expose the three heads of the deep digital flexor muscle. The tendons of all three heads fuse at the carpus to form a single tendon. This tendon is held in place in the carpal canal by the thick fibrous **flexor retinaculum.** The **carpal canal** is formed by the accessory carpal bone laterally, the joint capsule and carpal bones dorsally and the flexor retinaculum on the palmar surface. Cut this retinaculum medially and reflect it laterally to the accessory carpal bone to expose the deep digital flexor tendon. Identify the three heads of origin. Notice that the **humeral head** of this muscle is much larger than the other two heads and has several bellies. The **ulnar head** is larger than the radial and arises from the caudal border of the ulna. The **radial head** is small and comes from the medial border of the radius. Distal to the carpus the deep digital flexor tendon divides into five branches. Each branch goes to the palmar surface of the base of the distal phalanx of its respective digit. There is a synovial bursa under the humeral head at the elbow and a carpal synovial sheath in the carpal canal. Digital synovial sheaths extend from above the metacarpophalangeal joints to the insertion of the tendons on the distal phalanges of all the digits. The digital synovial sheaths, except that of the first digit, are common to the superficial as well as deep flexor tendons. Dissect the deep digital flexor tendon to its insertion on the distal phalanx of the third digit. It has already been exposed with the superficial digital flexor tendon at the metacarpophalangeal joint. Note the **annular digital ligaments** that support the deep digital flexor tendon proximal and distal to the palmar surface of the proximal interphalangeal joint (Fig. 26).

ORIGIN: Humeral head — the medial epicondyle of the humerus; ulnar head — the proximal three-fourths of the caudal border of the ulna; radial head — the middle third of the medial border of the radius.

INSERTION: The palmar surface of the base of the distal phalanx of each digit.

ACTION: To flex the digits.

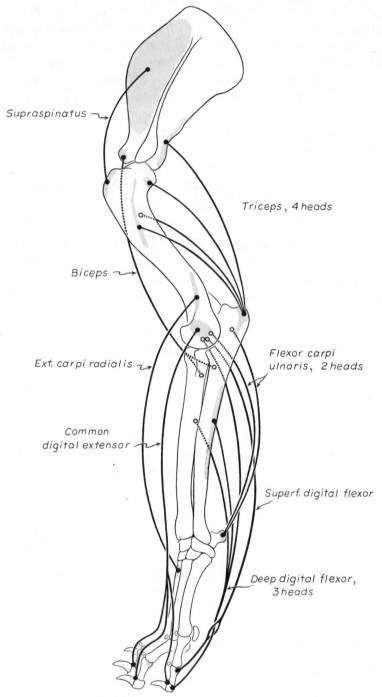

Supraspinatus

Triceps , 4 heads

Biceps

Ext. carpi radialis

Flexor carpi ulnaris, 2 heads

Common digital extensor

Superf. digital flexor

Deep digital flexor, 3 heads

Figure 27. Major extensors and flexors of forelimb.

6. The **pronator quadratus** (Figs. 14, 20, 23, 24) fills in the space between the radius and ulna. Spread the flexor muscles and observe the pronator. The fibers of this muscle run transversely between the ulna and radius.

ATTACHMENTS: The apposed surfaces of the radius and ulna.

ACTION: To pronate the paw.

MUSCLES OF THE FOREPAW

There are several special muscles of the digits. Only the **interossei** will be dissected. The four interossei are fleshy and similar in size and shape (Fig. 26). They lie deep to the deep digital flexor tendon and cover the palmar surfaces of the four main metacarpal bones. Transect the deep digital flexor tendon at the proximal end of the carpus and reflect it distally. Dissect the interosseous muscle of the third digit. Each muscle arises from the proximal end of its respective metacarpal bone and the carpal joint capsule. After a short course, each divides into two tendons which attach to the proximal end of the first phalanx. Imbedded in each tendon is a sesamoid bone which lies on the palmar surface of the meta-carpophalangeal joint. There are thus two **proximal sesamoids** at meta-carpophalangeal joints II, III, IV and V. A lesser tendon continues obliquely across the proximodorsal end on each side of the first phalanx and joins the common extensor tendon. The interosseous muscle is a flexor of the metacarpophalangeal (fetlock) joint. Complete exposure is facilitated by severing the tissues between the second and third, and third and fourth digits and metacarpal bones. Abduct the third digit to break the carpo-metacarpal joint and expose the interosseous muscle.

JOINTS OF THE THORACIC LIMB

The **shoulder** (Fig. 28) is a ball-and-socket joint between the glenoid cavity of the scapula and the head of the humerus. It is capable of movements in any direction but the chief movements it undergoes are flexion and extension. The shoulder joint capsule is a loose sleeve of synovial membrane and thin fibrous tissue which unites the scapula and humerus. The muscles which cross the joint help stabilize the articula-

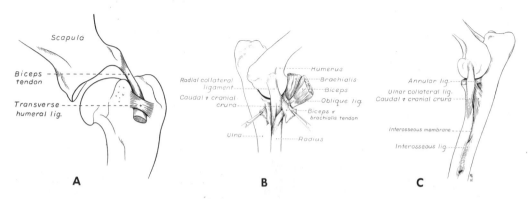

Figure 28.   *a. Shoulder, medial view.*
b. Left elbow, medial view.
c. Left elbow, lateral view.

(*b* and *c* from Miller et al., Anatomy of the Dog, 1964.)

tion. There is a collagenous thickening across the tendon of origin of the biceps at the intertubercular groove; this is the **transverse humeral ligament.**

The **elbow** is formed by the distal end of the humerus and the proximal end of the radius and trochlear notch of the ulna. To these should be added the proximal radio-ulnar articulation, although this is not weight bearing. The elbow joint capsule attaches to the articular margins; it extends distally a short distance between the radius and ulna. All compartments communicate with each other. Only the **lateral** and **medial collateral ligaments** are pronounced thickenings in the fibrous layer of the capsule. The biceps and brachialis tendons cover the distal portion of the medial collateral ligament. Clean these on your specimen.

The **interosseous ligament** is a condensation of collagenous tissue which unites the radius and ulna proximally.

The **carpal joint capsule** is usually tight-fitting as it extends as a sleeve from the distal ends of the radius and ulna to the metacarpus. It attaches to the carpal bones in its course across the joint. The compartments of the capsule are (1) between the radius and ulna and the proximal row of carpals, (2) between the two rows of carpal bones and (3) between the carpus and metacarpus. The cavity between the proximal row of carpals and the radius and ulna does not communicate with the other two cavities. No dissection is necessary.

The ligaments of the carpus differ from those of typical hinge joints in that both the dorsal and palmar surfaces are heavily reinforced by the fibrous layer of the joint capsule. On the dorsal surface of the joint the fibrous layer of the capsule contains grooves in which the extensor tendons lie. This layer is loose between the radius and ulna proximally and the proximal row of carpals distally, as most of the movement of the carpus takes place here. Cut through the joint capsule of the radiocarpal joint to observe its components and the degree of movement of the joint.

The **metacarpophalangeal, proximal interphalangeal** and **distal interphalangeal** joints are the three articulations of each main digit. Medial and lateral collateral ligaments support these joints.

The metacarpophalangeal joints each include two proximal sesamoids in the tendons of the interossei which articulate with the flexor surface of the metacarpal trochlea.

BONES OF THE PELVIC LIMB

The **pelvic girdle** or pelvis of the dog consists of two hip bones which are united with each other at the symphysis pelvis midventrally and the sacrum dorsally. Each hip bone or **os coxae** is formed by the fusion of three primary bones and the addition of a fourth in early life (Fig. 29). The largest and most cranial of these is the **ilium,** which articulates with the sacrum. The **ischium** is the most caudal, while the **pubis** is located ventrally cranial to the large obturator foramen. The **acetabulum,** a socket, is formed where the three bones meet. This receives the head of the femur in the formation of the hip. The small **acetabular bone,** which helps form

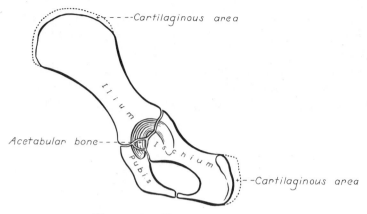

Figure 29. Hip bone of puppy.

the acetabulum, is incorporated with the ilium, ischium and pubis when these fuse (about the third month).

The **pelvic canal** is short ventrally but long dorsally. Its lateral wall is composed of the ilium, ischium and pubis. Dorsolateral to the skeletal part of the wall the pelvic canal is bounded by soft tissues. The **pelvic inlet** is limited laterally and ventrally by the arcuate line. Its dorsal boundary is the promontory of the sacrum. The **pelvic outlet** is bounded ventrally by the **ischiatic arch** (the ischiatic arch is formed by the concave caudal border of the two ischii), middorsally by the first caudal vertebra and laterally by the superficial gluteal muscle and the sacrotuberous ligament.

OS COXAE

1. The **ilium** (Figs. 29–31), a flat bone presenting two surfaces and three borders, forms the cranial half to three-fifths of the os coxae. It can be divided into a wide cranial part, which is concave laterally and is known as the **wing,** and a narrow, laterally-compressed caudal part, the **body.**

The cranial border is arciform and usually roughened and is more commonly known as the **iliac crest.** It is thin but gradually increases in thickness dorsally. The angle of junction of the iliac crest with the ventral border is known as the **cranial ventral iliac spine,** which provides a place of origin for both bellies of the sartorius and a part of the tensor fasciae latae.

On the ventral border a few centimeters from the cranial ventral iliac spine is the **caudal ventral iliac spine.** The **tuber coxae** is composed of the cranial and caudal ventral iliac spines and the intervening ventral border of the ilium. The rest of the ventral border is concave. It ends in the tuberosity for the rectus femoris just cranial to the acetabulum.

The dorsal border of the ilium is broad and massive. The junction of the dorsal border with the iliac crest forms an obtuse angle which is a rounded prominence, the **cranial dorsal iliac spine.** Caudal to the cranial dorsal iliac spine is the wide but blunt **caudal dorsal iliac spine.** The two spines and intervening bone occupy nearly half the length of the dorsal border of the ilium. These spines and the bone between them are known collectively as the **tuber sacrale.** The caudal half of the dorsal border is gently concave. It forms the **greater ischiatic notch** and also helps form the ischiatic spine which is dorsal to the acetabulum.

The external or **gluteal surface** (Fig. 30) of the wing of the ilium is nearly flat caudally and concave cranially where it is limited by the iliac crest. The dorsal part of this concave area is bounded by a heavy ridge. The gluteal surface is rough ventrocranially.

The internal or **sacropelvic surface** (Fig. 31) of the wing of the ilium presents a smooth, nearly flat area which gives origin to the longissimus and the quadratus lumborum muscles. The **auricular surface** is rough and articulates with a similar surface of the sacrum, forming the iliosacral joint. The **arcuate line** is located along the ventromedial edge of the sacropelvic surface of the body of the ilium and runs from the auricular surface to the iliopubic eminence of the pubis.

2. The **ischium** (Fig. 30) consists of tuberosity, body and ramus. It forms the caudal part of the os coxae and enters into the formation of the acetabulum, obturator foramen and symphysis pelvis. The **ischiatic tuberosity** is the thick caudolateral margin of the bone. The lateral angle of the tuber is enlarged and hooked; it furnishes attachment for the sacrotuberous ligament. The medial angle is rounded. The ventral surface fur-

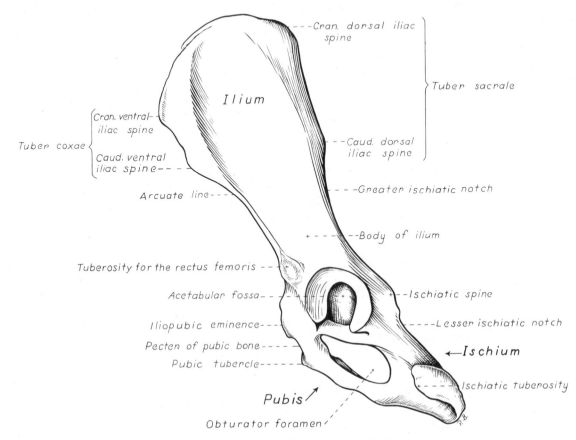

Figure 30. *Left hip bone, lateral view.*

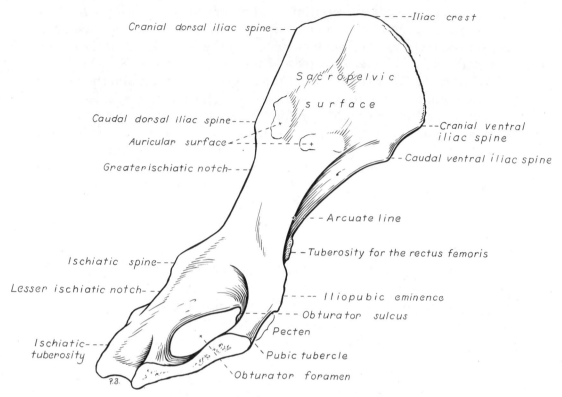

Figure 31. Left hip bone, medial view.

nishes origin for the biceps femoris, semitendinosus and semimembranosus. The crus of the penis and the muscle which surrounds it also attach to the ischiatic tuberosity.

The internal obturator arises from the dorsal surface of the ischium. The ventrolateral surface of the ischium (Fig. 30) is broad and expansive caudally, where it gives rise to the quadratus femoris, the external obturator and the adductor femoris. From the convex cranial part of this surface arise the gemelli.

The **ischiatic spine,** a rounded crest, is continuous from the ilium along the dorsal part of the body of the ischium. This is limited caudally by a series of low ridges produced by the tendon of the internal obturator. This area is known as the **lesser ischiatic notch** (Figs. 30, 31).

The **ramus** of the ischium is thin and wide, is bounded laterally by the obturator foramen, and blends caudally with the body of the ischium. It meets its fellow at the symphysis and is fused with the pubis cranially. The adductor femoris and the external obturator arise in part from its ventral surface; the internal obturator arises in part from its dorsal surface.

The **ischiatic arch** is formed by the caudal border of each ischium.

3. The **pubis** (Figs. 29-31) extends from the ilium and ischium laterally to the symphysis medially and consists of a body and two rami. The **body** is located cranial to the obturator foramen. The **cranial ramus**

extends from the body to the ilium and enters into the formation of the acetabulum. The **caudal ramus** fuses with the ischium at the middle of the pelvic symphysis. The ventral surface of the pubis serves as origin for the gracilis, the adductor femoris and the external obturator. The dorsal surface gives rise to a small part of the internal obturator and the levator ani. The **obturator sulcus,** a groove for the obturator nerve, is located at the cranial end of the obturator foramen and passes dorsally over the pelvic surface of the body of the bone. The cranial border of the pubis may be roughened to form the **pecten,** which extends from the iliopubic eminence to the pubic tubercle. The latter projects cranially from the pubis on the midline. The pectineus arises from the iliopubic eminence. The psoas minor is inserted on the tubercle dorsal to this. The rectus abdominis and adductor longus attach to the pecten.

The **acetabulum** is a cavity which receives the head of the femur. Its articular surface is semilunar and is composed of parts of the ilium, ischium and pubis and, in young animals, the acetabular bone (Fig. 29). The circumference of the articular surface is broken at the caudomedial part by the **acetabular notch.** The **acetabular fossa** is formed by the ischium and the acetabular bone. The ligament of the head of the femur attaches in this fossa. The fossa and the notch are the non-articular parts of the acetabulum. The two sides of the notch are connected by the transverse ligament.

The **obturator foramen** is closed in life by the obturator membrane and the external and internal obturator muscles which the membrane separates.

FEMUR

The femur (Figs. 33, 34), or thigh bone, is the largest bone in the body. The flexor angle of the hip is about 110 degrees. The flexor angle at the stifle is from 130 to 135 degrees.

The femur is a typical long bone with a cylindrical body and two expanded extremities. The proximal extremity presents on its medial side a smooth, nearly hemispherical **head,** most of which is articular, except for a small shallow fossa beginning near the middle of the head and usually extending to its caudomedial margin. This, the **fovea capitis femoris,** gives attachment to the ligament of the head of the femur. The head is attached to the medial part of the proximal extremity by the **neck** of the femur. The neck is distinct but short, and provides attachment for the joint capsule. The **greater trochanter,** the largest eminence of the proximal extremity, is located directly lateral to the head. To it attach the middle gluteal and deep gluteal. The **trochanteric fossa** is a deep cavity medial to the greater trochanter. The gemelli and the external and internal obturators insert in this fossa. The **lesser trochanter,** a pyramidal projection at the proximal end of the medial side of the body of the femur, serves for the insertion of the iliopsoas. A ridge of bone extends from the summit of the greater trochanter to the lesser trochanter. This, the **inter-**

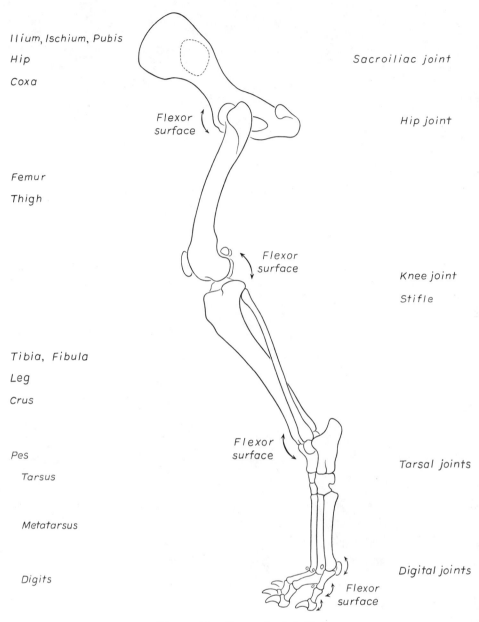

Ilium, Ischium, Pubis

Hip

Coxa

Sacroiliac joint

Flexor surface

Hip joint

Femur

Thigh

Flexor surface

Knee joint

Stifle

Tibia, Fibula

Leg

Crus

Flexor surface

Pes

Tarsus

Tarsal joints

Metatarsus

Digits

Digital joints

Flexor surface

Figure 32. Parts of pelvic limb.

trochanteric crest, represents the caudolateral boundary of the trochanteric fossa. The quadratus femoris inserts on the crest at the level of the lesser trochanter. The **third trochanter** is poorly developed. It appears at the base of the greater trochanter as a small rough area on which the superficial gluteal inserts. The third and lesser trochanters are located in about the same transverse plane. The vastus parts of the quadriceps femoris attach to the smooth proximal part of the femur.

The **body** of the femur is slightly convex cranially. Viewed cranially the body presents a smooth, rounded surface. The caudal or **rough surface** is limited by **medial** and **lateral lips.** The lips, closest together in the middle of the body, diverge as they approach each extremity. The

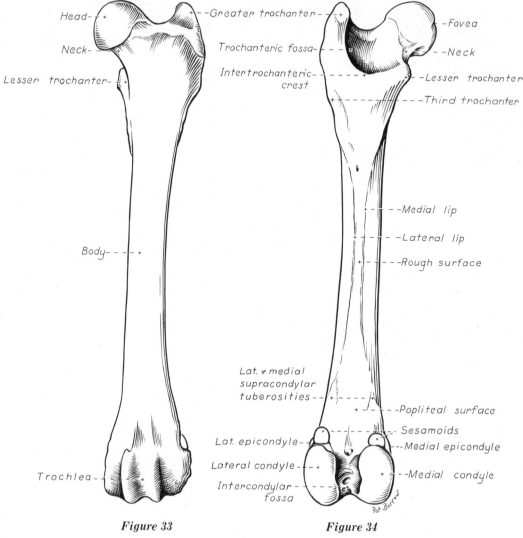

Figure 33

Left femur, cranial view.

Figure 34

Left femur, caudal view.

proximal part of the medial lip ends in the lesser trochanter, the distal part at the medial supracondylar tuberosity. The proximal part of the lateral lip ends in the third trochanter, the distal part at the lateral supracondylar tuberosity. The adductor inserts on most of the rough surface, while the pectineus sends a tendon to the distal part of the medial lip.

The distal extremity of the femur presents several articular surfaces. The **trochlea** is the smooth groove on the craniodistal part of the bone for articulation with the **patella.** The medial trochlear lip is usually thicker than the lateral. The patella (knee cap) is a sesamoid in the tendon of insertion of the large quadriceps femoris which extends the stifle. It aids in the protection of the tendon and the joint but its chief purpose is redirection of the tendon of insertion of the quadriceps. The trochlea of the femur is continuous with the condyles which articulate, both directly and through the menisci, with the tibia. The **medial** and **lateral condyles** are separated from each other by the **intercondylar fossa,** a deep, wide space. The two condyles are similar in shape and surface area. Each is convex transversely and longitudinally. At the depths of the intercondylar fossa, angular or circumscribed areas are present for the attachment of the cruciate ligaments. The two sesamoids located in the two tendons of origin of the gastrocnemius form medial and lateral facets caudodorsal to the respective condyles. The **popliteal surface** is a large, flat, triangular area on the caudal surface of the distal extremity proximal to the condyles and intercondylar fossa. Proximal to the facets at the proximal edge of the popliteal surface are located the **medial** and **lateral supracondylar tuberosities.** The gastrocnemius arises from both of these tuberosities. The superficial digital flexor arises from the lateral tuberosity. The **medial** and **lateral epicondyles** are rough areas on each side, proximal to the condyles. They serve for the attachment of the collateral ligaments of the stifle. The lateral epicondyle also gives rise to the popliteus. The tiny **extensor fossa** is located on the lateral epicondyle at the junction of the lateral condyle and the lateral lip of the trochlea; from it arises the long digital extensor. The semimembranosus is inserted just proximal to the medial epicondyle.

TIBIA

The tibia (Figs. 35, 36), or shin bone, has a proximal articular surface that flares out transversely and is also broad craniocaudally. It is wider than the distal end of the femur with which it articulates and is composed largely of two condyles. The **medial condyle** is separated from the **lateral condyle** by the **intercondylar eminence.** Both condyles include the articular areas on their proximal surfaces and the adjacent non-articular parts of the proximal extremity. The lateral condyle is particularly prominent. It possesses a facet on its lateral side for articulation with the head of the fibula and provides origin for part of the peroneus longus and cranial tibial muscles. The semimembranosus is inserted on the medial condyle. Two menisci, biconcave fibrocartilages, fill the space between the apposed condyles of the femur and tibia, making the joint congruent. The intercondylar eminence consists of two small, elongated tubercles, which form its highest part, and a shallow intermediate fossa. The **cranial intercondylar area** is a depression cranial to the eminence and in large part between the condyles. It affords attachment to the cranial parts of the menisci and the cranial cruciate ligament. The **caudal intercondylar area** occupies a place similar to that of the cranial area but caudal to the eminence. It provides attachment for the caudal parts of the menisci and the caudal cruciate ligament. The **popliteal notch** is caudal to the caudal intercondylar area and is located between the two condyles. The popliteus passes over the lateral condyle and through the notch. The **tibial tuberosity** is the large quadrangular process on the proximocranial surface of the tibia. The quadriceps femoris, the biceps femoris and the sartorius attach to this tuberosity by means of the patella and patellar ligament. The tibial tuberosity is continued distally by the cranial border of the

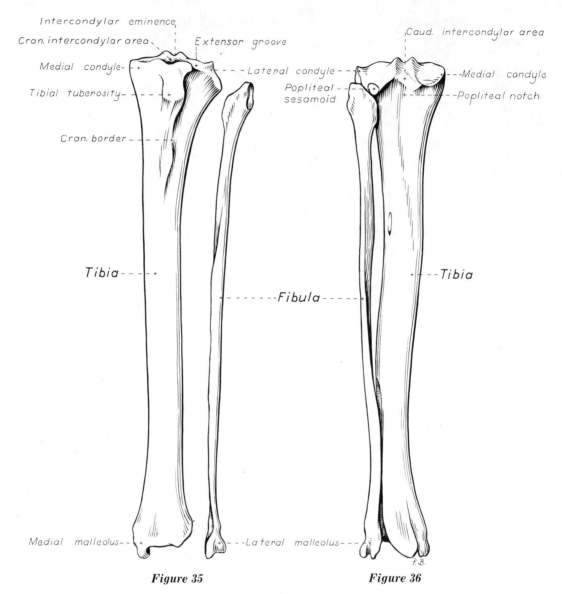

Figure 35

Left tibia and fibula, cranial view.

Figure 36

Articulated tibia and fibula, caudal view.

tibia. It inclines laterally on the shaft. The following muscles attach wholly or in part to the cranial border of the tibia: biceps femoris, semitendinosus, gracilis and sartorius. The **extensor groove** is a small, smooth groove located at the junction of the lateral condyle and the tibial tuberosity. The long digital extensor passes through it.

The **body** is triangular proximally, nearly cylindrical in the middle and four-sided distally. The semitendinosus and gracilis are inserted on the proximal medial surface. The proximal third of the caudal surface serves for the insertion of the popliteus medially and for the origins of the deep digital flexor laterally.

The distal extremity of the tibia is quadrilateral in cross section. The **tibial cochlea,** the articular surface, consists of two grooves which receive the ridges of the proximal trochlea of the talus. The medial part of the distal extremity of the tibia is the **medial malleolus.** The lateral surface of the distal extremity articulates with the fibula by a small facet. No muscles arise from the distal half of the tibia.

FIBULA

The fibula (Figs. 35, 36) has proximal and distal extremities and an intermediate body. The proximal extremity, or **head,** articulates with the lateral condyle of the tibia. The distal extremity, **lateral malleolus,** has two grooves. On its medial surface is a distinct facet for articulation with the distolateral surface of the tibia and with the talus.

TARSAL BONES

The **tarsus** (Fig. ·37), between the metatarsus and the leg, is comprised of seven tarsal bones and the related soft tissues. It is also called the hock. The bones are arranged in three irregular rows. The proximal

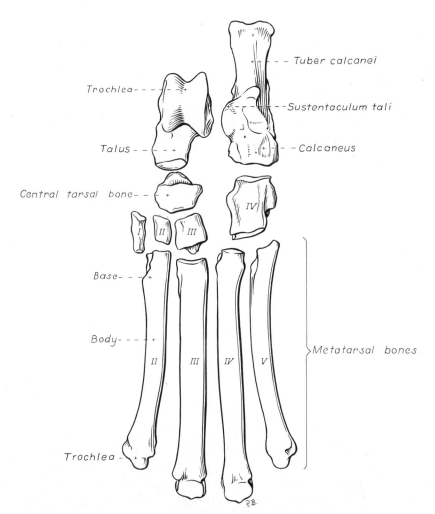

Figure 37. *Tarsal and metatarsal bones, dorsal view.*

row is composed of a long, laterally located **calcaneus** and a shorter, medially located **talus.** The tibia and fibula articulate with the talus. The distal row consists of four bones. Three small bones, the **first, second** and **third tarsal** bones, are located side by side and are separated from the proximal row by the **central tarsal** bone. The large **fourth tarsal** bone, which completes the distal row laterally, is as long as the combined lengths of the third and central tarsal bones against which it lies.

The **talus** has a trochlea on its proximal end for articulation with the tibial cochlea. The **tuber calcanei** is a traction process of the calcaneus that projects proximally and caudally. The extensor muscles of the hock insert on this process via the calcanean tendon. On the medial side of the calcaneus is a bony process, the **sustentaculum tali.** The tendon of the lateral head of the deep digital flexor glides over the plantar surface of this process. The fourth tarsal bone is grooved on the distal half of its lateral surface for the passage of the tendon of the peroneus longus.

METATARSAL BONES

The metatarsal bones (Fig. 37) resemble the metacarpal bones.

PHALANGES

The phalanges and sesamoids form the skeleton of the digit. Those of the hind paw are similar to those of the forepaw.

The first digit is frequently absent. When present, it is called a dewclaw and may vary from a fully developed digit articulating with a normal first metatarsal bone to a vestigial structure composed only of a terminal phalanx.

MUSCLES OF THE PELVIC LIMB

Remove the skin from the caudal part of the left half of the trunk, the rump and the thigh. Continue the midventral incision from the umbilicus to the root of the tail. In making this incision closely circle the external genital parts and the anus. Extend an incision distally on the medial surface of the left thigh to the tarsus. Encircle the tarsus with a skin incision. First reflect the skin from the medial surface of the thigh and then, starting at the tarsus, reflect the whole flap of skin from the lateral surface of the leg, stifle, thigh, rump and abdomen to the middorsal line. The cutaneus trunci may be removed with the skin since it is more intimately attached to the skin than to the underlying structures.

There are superficial and deep fasciae of the pelvic limb. They cannot always be separated and, in general, the deep one is the stronger.

The **superficial fascia of the trunk** (Fig. 38) continues dorsally on the rump as the **superficial gluteal fascia.** In obese specimens

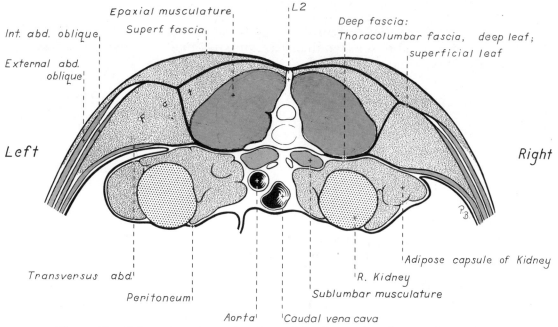

Figure 38. Schematic cross-section through lumbar region showing fascial layers.

much fat is present between this fascia and the deep fascia. The cutaneus trunci arises as an aponeurosis from the superficial gluteal fascia. It may be examined in the skin just reflected and in the previous dissection. Its fibers run cranioventrally to the caudal part of the axilla. Its most ventral border lies in the fold of the flank; dorsally it is separated from its fellow by a narrow band of fascia over the loin and rump. The superficial gluteal fascia passes to the tail as the **superficial caudal fascia** and continues distally on the limb as the superficial fascia of its respective parts. Remove the fat-laden areolar tissue and the superficial fascia covering the caudal part of the trunk and rump. Do not cut into the glistening deep fascia underneath.

A thick deep fascia covers the dorsal muscles of the loin, rump and tail. On the dorsocaudal part of the trunk this is the **thoracolumbar fascia,** a region of the **deep fascia of the trunk** (Fig. 38). It is continued caudally at the iliac crest by the **deep gluteal fascia.** This distinct glistening fascia covers the muscles of the rump and serves partially as origin for the middle gluteal. The deep gluteal fascia is continued caudally on the tail by the **deep caudal fascia.** The latter follows the irregularities of the caudal muscles and closely binds these to the vertebrae of the tail. Distally, the deep gluteal fascia blends with the fascia of the thigh, where it becomes the **medial** and **lateral femoral fasciae.** The medial fascia is thin whereas the lateral femoral fascia, the **fascia lata,** is thick and serves as an aponeurotic insertion for thigh muscles. The femoral fascia is continued on the leg as the **crural fascia.** Clean but do not cut any of the deep fascia until instructed to do so.

CAUDAL MUSCLES OF THE THIGH

This group consists of four muscles: biceps femoris, laterally; caudal crural abductor and semitendinosus, caudally; and semimembranosus, medially.

1. The **biceps femoris** (Figs. 39, 42, 48, 53, 54) is the longest and widest of the muscles of the thigh. The bulk of its fibers runs craniodistally, although caudally there are fibers which run directly distally. Cranially it inserts by means of the fascia lata and crural fascia. Carefully clean the caudal border and the adjacent surface of the muscle. The lymph node lying in the fat at its caudal border directly caudal to the stifle is the **popliteal lymph node.** Do not cut the cranially lying fascia lata which serves as part of its insertion. Caudally a strand of heavy fascia runs to the tuber calcanei and helps to form the **common calcanean tendon.** Transect the biceps. The ischiatic nerve is interposed between the biceps and the underlying muscles. Observe the insertions of the biceps femoris.

ORIGIN: The sacrotuberous ligament and the ischiatic tuberosity.

INSERTION: By means of the fascia lata and crural fascia to the patella, patellar ligament and cranial border of the tibia; by means of the crural fascia to the subcutaneous part of the tibial body; the tuber calcanei.

ACTION: To extend the hip, stifle and hock. The caudal part of the muscle flexes the stifle.

2. The **semitendinosus** (Figs. 39–44, 48, 53, 54), near its origin, lies between the biceps femoris and semimembranosus. Near its insertion it lies on the medial head of the gastrocnemius and is covered by the gracilis. It is nearly as wide as it is thick and extends principally from the ischiatic tuberosity to the tibial body. By means of the distal continuation of the crural fascia it also attaches to the tuber calcanei. Completely free this muscle from surrounding structures but do not transect it.

ORIGIN: The ischiatic tuberosity.

INSERTION: The medial surface of the body of the tibia and the tuber calcanei by means of the crural fascia.

ACTION: To extend the hip, flex the stifle and extend the hock.

3. The **semimembranosus** (Figs. 40-44, 46, 50) is greater in cross-sectional area than the semitendinosus, but is not as long. It is wedged between the semitendinosus and biceps laterally and the gracilis and adductor medially. It has two bellies of nearly equal size. This short but strong muscle extends from the ischiatic tuberosity to the medial side of the distal end of the femur and the proximal end of the tibia. The insertions may be seen more adequately after the sartorius and gracilis muscles have been dissected.

ORIGIN: The ischiatic tuberosity.

INSERTION: The medial lip of the rough surface of the femur and the medial condyle of the tibia.

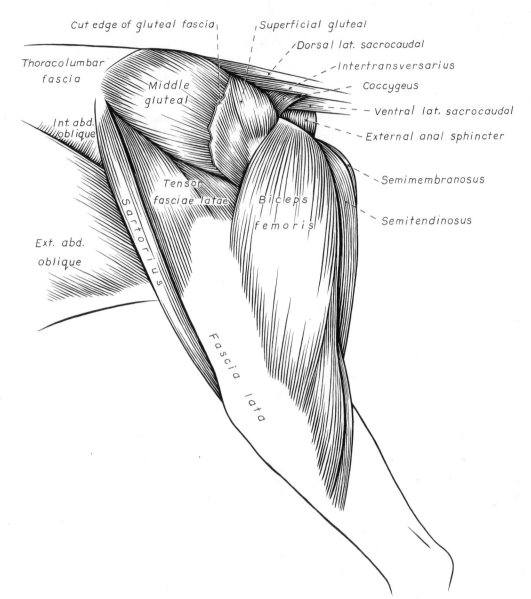

Figure 39. Superficial muscles of pelvic limb, lateral view.

ACTION: To extend the hip. The part which attaches to the femur extends the stifle; the part which attaches to the tibia flexes or extends the stifle depending upon the position of the limb.

MEDIAL MUSCLES OF THE THIGH

1. The **sartorius** (Figs. 39-43, 48) consists of two straplike parts which lie on the craniomedial surface of the thigh. They extend from the ilium to the tibia. The **cranial part** forms the cranial contour of the thigh and may be nearly a centimeter thick at this place. The **caudal part** is on the medial side of the leg and is thinner and wider than the cranial part.

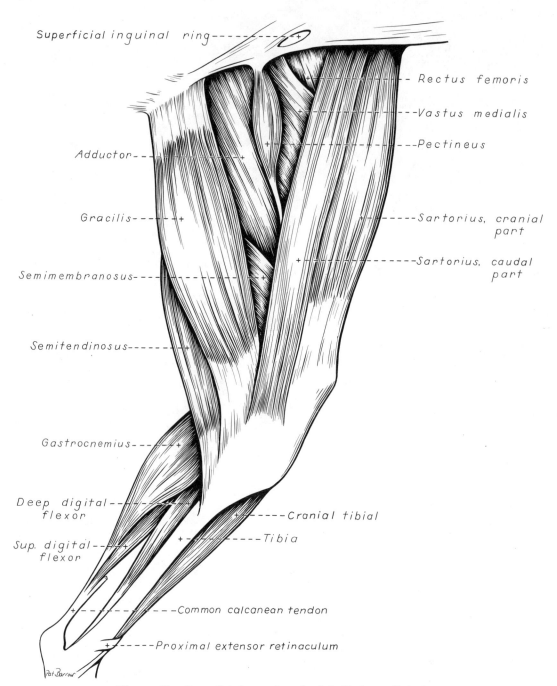

Figure 40. *Superficial muscles of pelvic limb, medial view.*

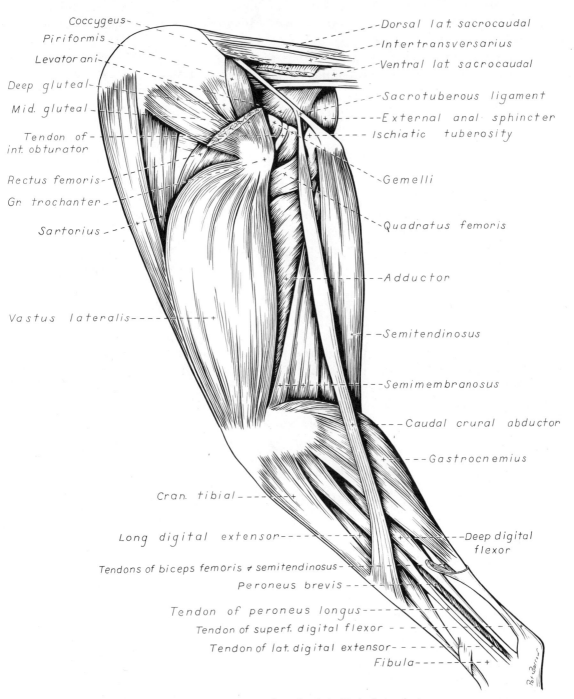

Figure 41. *Deep muscles of pelvic limb, lateral view.*

Both muscle parts lie predominantly on the medial side of the large quadriceps femoris. Transect both parts and reflect the distal parts to their insertions.

ORIGIN: Cranial part—the crest of the ilium and the thoracolumbar fascia; caudal part—the cranial ventral iliac spine and the adjacent ventral border of the ilium.

INSERTION: Cranial part—the patella, in common with the rectus femoris of the quadriceps; caudal part—the cranial border of the tibia, in common with the gracilis.

ACTION: To flex the hip. The cranial part extends the stifle; the caudal part flexes the stifle.

2. The **gracilis** (Figs. 40, 42, 43, 48) arises from the **subpelvic tendon,** a thick, flat tendon attached ventrally to the symphysis pelvis. The aponeurosis of the gracilis covers the adductor. Transect the gracilis through its aponeurotic origin. Reflect it distally and observe its insertion as well as that of the semimembranosus.

ORIGIN: The pelvic symphysis by means of the subpelvic tendon.

INSERTION: The cranial border of the tibia and, with the semitendinosus, the tuber calcanei.

ACTION: To adduct the limb, flex the stifle and extend the hip and hock.

The **femoral triangle** is the shallow triangular space through which the femoral vessels run to and from the pelvic limb. It is located on the proximal medial surface of the thigh with its base at the abdominal wall. The triangle lies between the caudal belly of the sartorius cranially and the pectineus and adductor caudally. The iliopsoas forms the proximal lateral part of the triangle. The vastus medialis forms the distal lateral part. This triangle contains, among other structures, the femoral artery and vein. Remove the medial femoral fascia and adipose tissue which covers and fills in around the femoral artery and vein. Notice that the vein lies caudal to the artery. The pulse is usually taken from the femoral artery here.

3. The **pectineus** (Figs. 40, 42-44, 46) is a small, spindle-shaped muscle which, with the adductor, belongs to the deep medial muscles of the thigh. It lies in large part between the adductor caudally and the vastus medialis cranially. It arises from the iliopubic eminence and the **cranial pubic ligament.** The latter is composed of transverse fibers that connect the pecten of one side with that of the other side. The tendon of insertion of the pectineus lies interposed between the adductor and vastus medialis. By blunt dissection with the handle of the scalpel isolate the tendon of insertion. It inserts on the caudomedial surface of the distal end of the femur. Transect the pectineus in the middle of its belly.

ORIGIN: The cranial pubic ligament and the iliopubic eminence.

INSERTION: The distal end of the medial lip of the rough face of the femur.

ACTION: To adduct the limb and rotate the leg outward.

4. The **adductor** (Figs. 40-42, 44-46) is divisible into two muscles. One of these may be seen without dissection as the large pyramidal muscle which is compressed between the semimembranosus and pectineus. This is the **adductor magnus et brevis.** The smaller **adductor longus** (Fig. 45) is fusiform and much like the pectineus, although shorter and caudolateral to it. It may be found deep (lateral) and slightly cranial to the proximal end of the adductor magnus et brevis. By blunt dissection separate the magnus et brevis from the longus. Transect the great adductor at its origin. Notice that the long adductor is completely covered by the adductor magnus et brevis and arises from the pubic tubercle and inserts with the great adductor.

ORIGIN: Adductor magnus et brevis—the entire pelvic symphysis by means of the subpelvic tendon, the adjacent part of the ischiatic arch and the ventral surface of the pubis and ischium; adductor longus—the pubic tubercle.

INSERTION: Adductor magnus et brevis—the entire lateral lip of the rough face of the femur; adductor longus—the proximal end of the lateral lip of the rough surface, near the third trochanter.

ACTION: To adduct the limb and extend the hip.

LATERAL MUSCLES OF THE RUMP

1. The **tensor fasciae latae** (Figs. 39, 42) is a triangular muscle which attaches proximally to the tuber coxae. It lies between the sartorius cranially, the middle gluteal caudally and the quadriceps distomedially. Part of its caudal surface is attached to the middle gluteal near its origin. The muscle can be divided into two portions. The cranial, more superficial portion is inserted on the lateral femoral fascia which radiates over the quadriceps and blends with the fascial insertion of the biceps femoris. The deeper caudal portion is inserted on a layer of lateral femoral fascia that runs deep to the biceps toward the stifle on the lateral surface of the vastus lateralis. Transect the tensor fasciae latae across its middle and reflect the two halves toward their attachments.

ORIGIN: The tuber coxae and adjacent part of the ilium; the aponeurosis of the middle gluteal muscle.

INSERTION: The lateral femoral fascia.

ACTION: To tense the lateral femoral fascia, flex the hip and extend the stifle.

2. The **superficial gluteal** (Figs. 39, 42, 46) is small and lies caudal to the middle gluteal. Its fibers run distally, from the gluteal fascia which covers the middle gluteal, the sacrum and the first caudal vertebra, as far as the greater trochanter of the femur, where they converge slightly before forming an aponeurosis which runs under the biceps to the third trochanter. Clean the superficial gluteal and transect it 1 cm. from the beginning of its aponeurosis of insertion. Do not transect the caudally lying **sacrotuberous ligament.** This is a collagenous band which runs from the sacrum to the lateral angle of the ischiatic tuberosity. Notice that the superficial gluteal

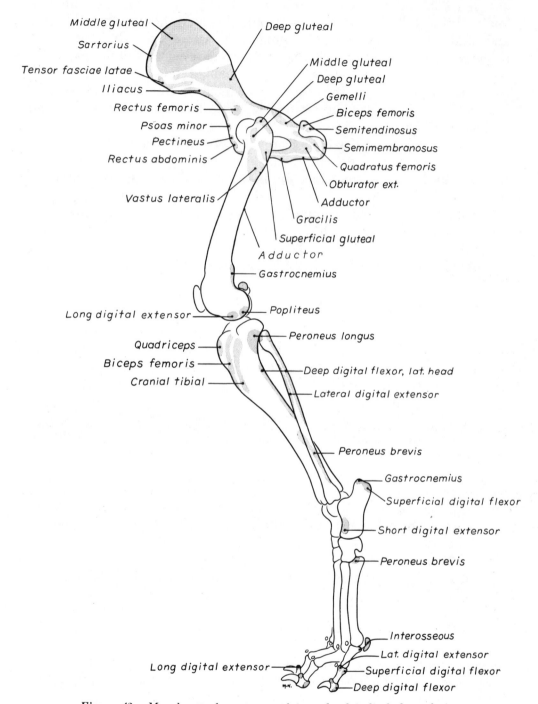

Figure 42. *Muscle attachments on pelvis and pelvic limb, lateral view.*

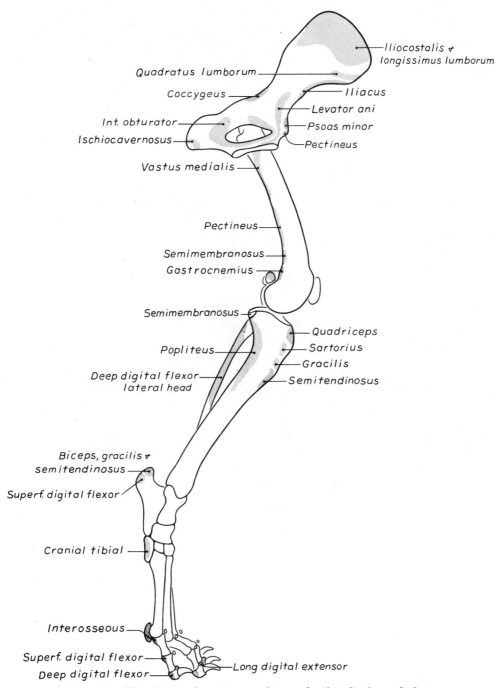

Figure 43. *Muscle attachments on pelvis and pelvic limb, medial view.*

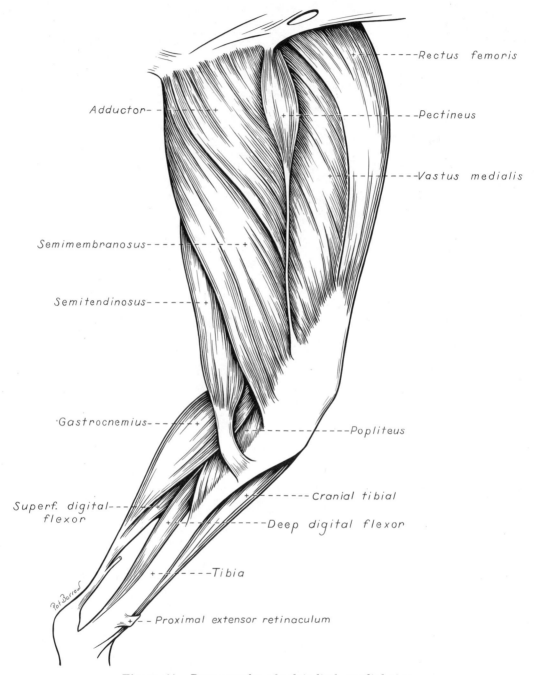

Figure 44. *Deep muscles of pelvic limb, medial view.*

arises from the proximal half of this ligament. Sever the deep gluteal fascia 1 cm. cranial to its junction with the muscle fibers of the superficial gluteal.

ORIGIN: The lateral border of the sacrum and the first caudal vertebra, partly by means of the sacrotuberous ligament; the cranial dorsal iliac spine by means of the gluteal fascia.

INSERTION: The third trochanter.

ACTION: To extend the hip and abduct the leg.

3. The **middle gluteal** (Figs. 39, 42, 46) is a large, ovoid muscle which lies between the tensor fasciae latae and the superficial gluteal. Clean the surface of the muscle by reflecting the gluteal fascia dorsocranial to the iliac crest. This reveals that the fibers of the muscle for the most part parallel its long axis. Carefully separate the cranioventral part of the middle gluteal from the underlying deep gluteal. The entire caudodorsal border of the middle gluteal is covered by the superficial gluteal. The deep caudal portion of the middle gluteal is readily separated from the main muscle mass. Starting at the middle of the cranioventral border of the middle gluteal, transect the entire muscle and reflect the distal half toward its insertion on the apex of the greater trochanter of the femur.

ORIGIN: The crest and gluteal surface of the ilium.

INSERTION: The greater trochanter.

ACTION: To extend and abduct the hip.

4. The **deep gluteal** (Figs. 41, 42, 47) is fan-shaped and completely covered by the middle gluteal. Superficially it is covered by an aponeurosis whose fibers converge to insert on the cranial face of the greater trochanter.

ORIGIN: The body of the ilium; the ischiatic spine.

INSERTION: The cranial aspect of the greater trochanter.

ACTION: To extend and abduct the hip.

CAUDAL HIP MUSCLES

The four muscles of this group are important because of their proximity to the hip. They lie caudal to the hip and extend from the inner and outer surfaces of the ischium to the femur.

1. The **internal obturator** (Figs. 41, 43, 46) is fan-shaped, with its muscle fibers converging toward the lesser ischiatic notch. The body of the muscle may be exposed at its origin on the dorsal surface of the ischium by removing the loose fat and the fascia caudomedial to the sacrotuberous ligament. The most caudal fibers of the internal obturator run craniolaterally toward the lesser ischiatic notch, where the tendon of the muscle begins. The tendon of the internal obturator passes over the lesser ischiatic notch ventral to the sacrotuberous ligament. Transect the sacrotuberous ligament and reflect the adjacent soft tissues to expose the tendon of insertion of the internal obturator muscle running to the tro-

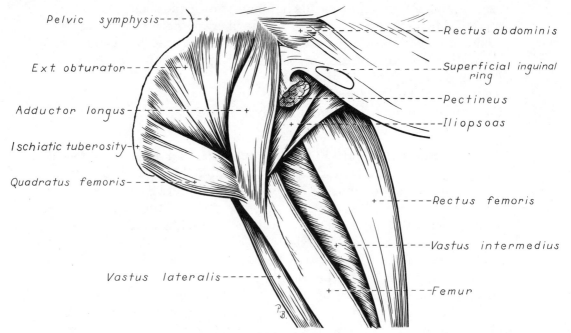

Figure 45. Deep muscles medial to hip.

chanteric fossa. Transect the tendon of the internal obturator as it crosses the gemelli and reflect it to observe the bursa which lies between the tendon and the lesser ischiatic notch.

ORIGIN: The symphysis pelvis and the dorsal surface of the ischium.
INSERTION: The trochanteric fossa of the femur.
ACTION: To rotate the pelvic limb outward at the hip.

2. The **gemelli** (Figs. 41, 42, 46), two muscles fused together, lie under the tendon of the internal obturator. They are interposed between the quadratus femoris and external obturator distally and the deep gluteal proximally. The gemelli are deeply grooved by the tendon of the internal obturator so that their edges overlap this tendon.

ORIGIN: The dorsolateral surface of the ischium.
INSERTION: The trochanteric fossa.
ACTION: To rotate the pelvic limb outward at the hip.

3. The **quadratus femoris** (Figs. 41, 42, 45, 46) is short and thick. It lies under the biceps, where it is interposed between the adductor distally and the external obturator and gemelli proximally. Its fibers are at right angles to the long axis of the thigh. It should be examined from both the medial and lateral sides. The dorsal border of the quadratus femoris lies closely applied to the ventral border of the gemelli.

ORIGIN: The ventral surface of the caudal part of the ischium.
INSERTION: Just distal to the trochanteric fossa.
ACTION: To extend and rotate the hip outward.

4. The **external obturator** (Figs. 42, 45, 46) is fan-shaped and arises

on the ventral surface of the pubis and ischium. It covers the obturator foramen. Its caudal border is covered by the quadratus femoris, while its cranial border is hidden by the adductor longus. Transect the adductor longus and follow the external obturator to its insertion.

ORIGIN: The ventral surface of the pubis and ischium.

INSERTION: The trochanteric fossa.

ACTION: To rotate the hip outward.

CRANIAL MUSCLES OF THE THIGH

1. The **quadriceps femoris** (Figs. 40-48) is divided into four heads of origin which are fused distally. It arises from the femur and the ilium and is inserted on the tibial tuberosity. The patella lies in the tendon of insertion. This muscle is the most powerful extensor of the stifle.

The **rectus femoris** is the most cranial component of the quadriceps femoris. Proximally it is circular in cross-section and passes between the vastus medialis and the vastus lateralis. Uncover the rectus femoris near its origin. Transect and reflect the proximal part. The rectus arises from a tuberosity on the ilium cranial to the acetabulum and inserts on the tibial tuberosity. It is a flexor of the hip.

The patella is intercalated in the strong tendon of insertion of the

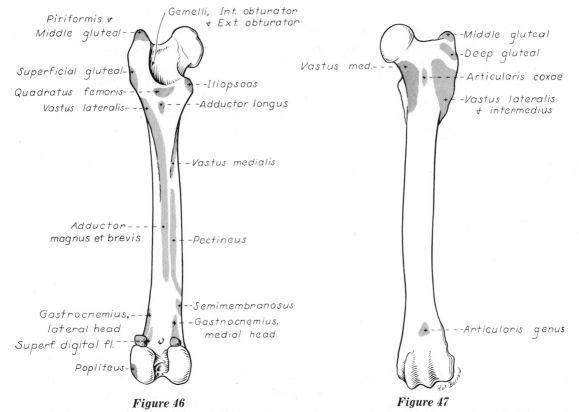

Figure 46

Femur with muscle attachments, caudal view.

Figure 47

Femur with muscle attachments, cranial view.

quadriceps. The patellar ligament, which extends from the patella to the tibial tuberosity, is but the distal end of the tendon of insertion of the quadriceps.

The **vastus lateralis** lies lateral and caudal to the rectus femoris, to which it is fused distally. The vastus lateralis is partly separated from the vastus intermedius by a scantly developed intermuscular septum. Notice that the vastus lateralis arises from the proximal part of the lateral lip of the rough face of the femur. Detach its caudal border from the femur. It is inserted with the rectus femoris on the tibial tuberosity.

The **vastus intermedius** lies directly on the smooth cranial surface of the femur and is quite intimately fused with the other two vasti. It arises with the vastus lateralis which covers it from the lateral side of the proximal end of the cranial surface of the femur. It inserts on the tibial tuberosity with the other members of the group.

The **vastus medialis** arises from the medial side of the proximal end of the cranial surface of the femur and the proximal end of the medial lip of the rough surface. It is inserted with the other heads of the quadriceps on the tibial tuberosity.

2. The **iliopsoas** (Fig. 45), a sublumbar muscle, is now visible at its insertion on the lesser trochanter of the femur. This end of the muscle lies between the pectineus medially and the rectus femoris laterally. The iliopsoas represents a fusion of the psoas major and iliacus muscles. The **psoas major** arises from transverse processes and bodies of lumbar vertebrae. It passes caudally and ventrally under the cranioventral aspect of the ilium, where it joins the iliacus. The **iliacus** arises from the smooth ventral surface of the ilium between the arcuate line and the lateral border of the ilium. The two muscle masses continue caudoventrally (as the iliopsoas) to their conjoined insertion on the lesser trochanter. The action of the iliopsoas is to flex the hip and the lumbar vertebral column.

MUSCLES OF THE LEG (CRUS)

The muscles of the leg or crus, the region between the stifle and hock, are divided into craniolateral and caudal groups. Remove the skin that remains on the distal part of the pelvic limb to the level of the proximal interphalangeal joints.

The **superficial crural, tarsal, metatarsal** and **digital fasciae** are similar to the superficial fasciae of the corresponding regions of the forelimb. Each is attached to the bones, where these are subcutaneous. The tarsal fascia extends into the metatarsal and digital pads and closely joins these pads with the overlying skeletal and ligamentous parts.

The medial and lateral femoral fasciae blend over the stifle and are continued distally in the leg as the **deep crural fascia.** The deep crural fascia covers the muscles of the leg and the free-lying surfaces of the crural skeleton. It sends intermuscular septae between the muscles to attach to the bone. Laterally the fibers of the caudal branch of the biceps femoris radiate into it. Caudally the crural fascia contributes to the common calcanean tendon.

Just proximal to the flexor surface of the tarsus the deep crural fascia is thickened to form an oblique band of about 0.5 cm., the **proximal extensor retinaculum.** As it stretches obliquely from the distal third of the fibula to the medial malleolus of the tibia it binds down the tendons of the long digital extensor and the cranial tibial muscles.

The deep crural fascia decreases in thickness as it passes over the tarsus and becomes the deep tarsal fascia. A distinct band of the deep tarsal fascia forms the **distal extensor retinaculum.** It attaches to the dorsal surface of the distal third of the calcaneus and holds the tendon of the long digital extensor in position. Observe this as the long digital extensor is dissected.

Make an incision through the cranial crural fascia and reflect it to the common calcanean tendon and the tibia. Do not remove the retinacula.

Craniolateral Muscles of the Leg

1. The **cranial tibial** (Figs. 42–44, 48, 52–54) is the most cranial muscle of this group. Its medial margin is in contact with the tibia. It

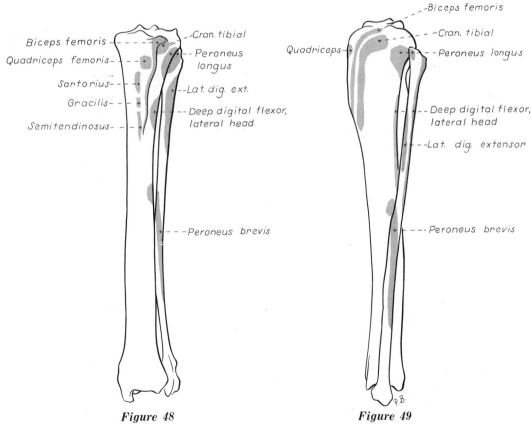

Figure 48

Tibia and fibula with muscle attachments, cranial view.

Figure 49

Tibia and fibula with muscle attachments, lateral view.

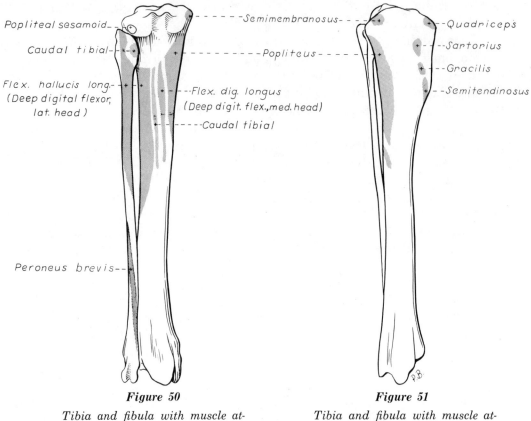

Figure 50

Tibia and fibula with muscle attachments, caudal view.

Figure 51

Tibia and fibula with muscle attachments, medial view.

arises from the cranial border and the adjacent proximal articular margin of the tibia. Its tendon inserts on the proximal plantar surface of the first and second metatarsals. The tendon of the cranial tibial runs under the proximal extensor retinaculum and is provided with a synovial sheath over most of the flexor surface of the tarsus.

ORIGIN: The muscular groove and the adjacent articular margin of the tibia; the lateral edge of the cranial tibial border.

INSERTION: The proximal plantar surface of metatarsals I and II.

ACTION: To rotate the foot laterally and to flex the tarsus.

2. The **long digital extensor** (Figs. 41–43, 52–54) is a spindle-shaped muscle that is partly covered by the cranial tibial medially and the peroneus longus laterally. Expose the muscle and its tendon of origin from the extensor fossa of the femur. The tendon runs over the articular margin of the tibia in the muscular groove and is lubricated by an extension of the stifle joint capsule. Observe the four tendons of insertion. As the tendons pass over the tarsus they are surrounded by a synovial sheath and are held in place by the proximal and distal extensor retinacula. In the metatarsus the four tendons diverge toward their respective digits.

ORIGIN: The extensor fossa of the femur.

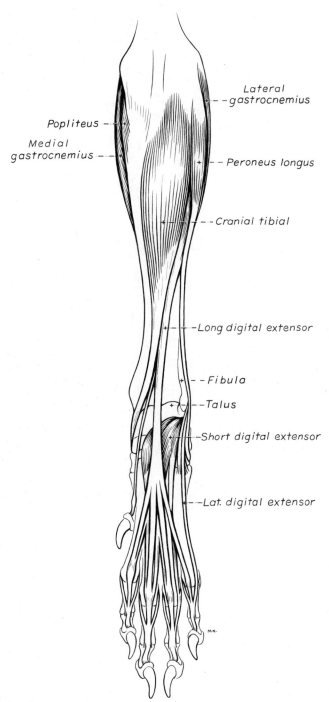

Popliteus

Medial gastrocnemius

Lateral gastrocnemius

Peroneus longus

Cranial tibial

Long digital extensor

Fibula

Talus

Short digital extensor

Lat. digital extensor

Figure 52. *Muscles of left pelvic limb, cranial view.*

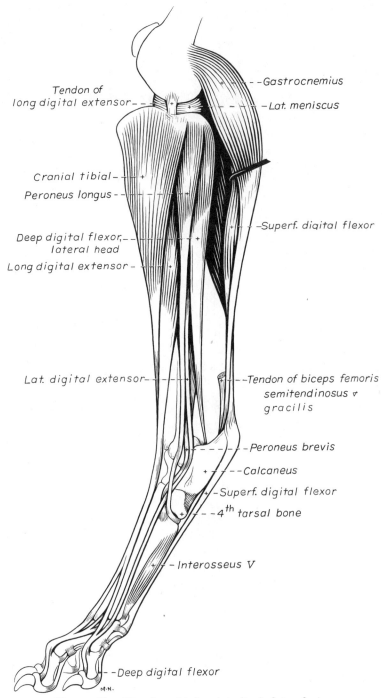

Figure 53.　*Muscles of left pelvic limb, lateral view.*

INSERTION: The extensor processes of the third phalanges of digits II, III, IV and V.

ACTION: To extend the digits and flex the tarsus.

3. The **peroneus longus** (Figs. 41, 42, 48, 52–54) lies just caudal to the long digital extensor, where a triangular portion of its short belly lies directly under the crural fascia. It is a short, thick, wedged-shaped muscle which lies in large part cranial to the fibula. It arises from the lateral ligament of the stifle and the adjacent parts of the tibia and fibula. Its stout tendon has a long synovial sheath which begins at a plane through the proximal extensor retinaculum and extends to its insertion on the fourth tarsal and the plantar surfaces of all the metatarsals. Open the synovial sheath at its proximal end. Preserving all ligaments and tendons which lie superficial to the tendon, trace it to its insertion on the fourth tarsal. Do not dissect the tendon beyond this attachment. Notice that it lies in a sulcus of the lateral malleolus of the fibula and that at the distal end of the tarsus it makes nearly a right angle as it turns caudally and medially around a groove in the fourth tarsal bone.

ORIGIN: The lateral condyle of the tibia, the proximal end of the fibula and the lateral epicondyle of the femur by means of the lateral ligament of the stifle.

INSERTION: The fourth tarsal bone; the plantar aspect of the proximal ends of metatarsals I, II and V.

ACTION: To flex the tarsus and turn the plantar surface of the paw laterally.

The lateral digital extensor and the peroneus brevis (Fig. 53) are located beneath the peroneus longus on the lateral aspect of the leg and need not be dissected.

Caudal Muscles of the Leg

1. The **gastrocnemius** (Figs. 40-44, 46, 52-54) is the main component of the common calcanean tendon. Notice that the tendon of the gastrocnemius passes deep to that of the superficial digital flexor. Near the middle of the leg the tendon of the gastrocnemius is thus subcutaneous. Distally it courses craniolateral to the superficial digital flexor tendon so that near its insertion on the calcaneal tuberosity it is entirely hidden by the flexor tendon. From the lateral side carefully dissect between the two tendons. By blunt dissection separate the two heads of the gastrocnemius from each other and that of the superficial digital flexor. At its origin the superficial digital flexor is closely bound to the lateral head of the gastrocnemius. Transect the gastrocnemius 4 cm. from its origin and separate its two heads. Observe: that the muscle, by two separable heads, arises from the medial and lateral supracondylar tuberosities of the femur; that its deep surface is grooved by the body of the superficial digital flexor; and that each head has intercalated in its tendon of origin a **sesamoid** which articulates with its respective condyle of the femur. Cut under these so that their identities are unmistakable. These sesamoids (Figs. 34, 36) are located on the caudal side of the stifle.

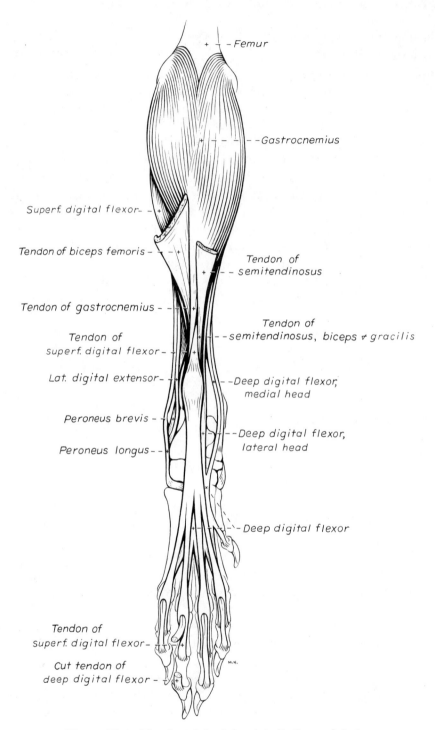

Figure 54a. *Muscles of the left pelvic limb, caudal view.*

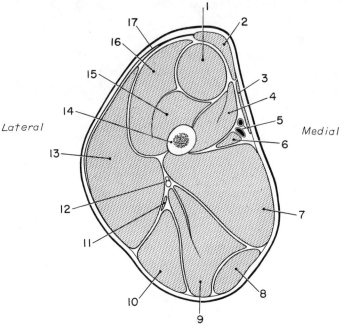

Figure 54b. Cross-section of thigh.

1. Rectus femoris
2. Sartorius, cranial part
3. Sartorius, caudal part
4. Vastus medialis
5. Femoral a. and v.
6. Pectineus
7. Adductor
8. Gracilis
9. Semimembranosus
10. Semitendinosus
11. Caudal crural abductor
12. Ischiatic n.
13. Biceps femoris
14. Femur
15. Vastus intermedius
16. Vastus lateralis
17. Fascia lata

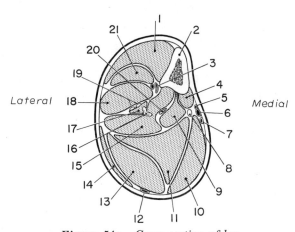

Figure 54c. Cross-section of leg.

1. Cranial tibial
2. Tibia
3. Cranial tibial a. and v.
4. Popliteus
5. Saphenous a., cranial branch
6. Medial saphenous v.
7. Saphenous a., caudal branch
8. Tibial n.
9. Deep digital flexor, medial head
10. Gastrocnemius, medial head
11. Superficial digital flexor
12. Lateral saphenous v.
13. Gastrocnemius, lateral head
14. Biceps femoris
15. Deep digital flexor, lat. head
16. Fibula
17. Lateral digital extensor
18. Peroneus longus
19. Peroneal nerves
20. Caudal tibial
21. Long digital extensor

In addition to these in the heads of origin of the gastrocnemius, a third sesamoid is present in the tendon of the popliteus as it rides over the lateral tibial condyle. This tendon will be dissected later.

ORIGIN: The medial and lateral supracondylar tuberosities of the femur.

INSERTION: The tuber calcanei.

ACTION: To extend the tarsus and flex the stifle.

2. The **superficial digital flexor** (Figs. 40, 42-44, 46, 53, 54) is a spindle-shaped muscle which arises from the lateral supracondylar tuberosity of the femur with the lateral head of the gastrocnemius. Its deep surface is in apposition to the deep digital flexor and the popliteus while its other surfaces are largely covered by the gastrocnemius. Proximal to the calcaneal process its tendon twists across the medial surface of the gastrocnemius. Farther distally the tendon widens, caps the tuber calcanei and attaches on each side. Transect the superficial digital flexor 1 cm. distal to the transection of the gastrocnemius. Make a sagittal incision through the superficial digital flexor tendon from the place where it gains the plantar side of the gastrocnemius to the tuber calcanei. Continue the incision distal to the tuber calcanei for an equal distance. Observe the large **calcaneal bursa** of the superficial digital flexor which lies under its tendon as it crosses the tuber calcanei. Notice the rather distinct medial and lateral attachments of the tendon on the tuber calcanei. Opposite the distal plantar surface of the tarsus the tendon bifurcates; each of these branches in turn bifurcates, thus forming four tendons of nearly equal size. Each tendon is disposed in its digit as the corresponding tendon in the forepaw. They need not be dissected.

ORIGIN: The lateral supracondylar tuberosity of the femur.

INSERTION: The tuber calcanei and the bases of the second phalanges of digits II, III, IV and V.

ACTION: To flex the first two digital joints of the four principal digits; flex the stifle; extend the tarsus.

3. The **deep digital flexor** (Figs. 40-44, 50, 53, 54) and the popliteus are the principal muscles yet to be dissected on the caudal surface of the leg. The separation between them runs distomedially. This division starts at the lateral tibial condyle and extends to the proximal third of the tibia on the medial side of the leg. The medial and lateral heads of the deep digital flexor are now exposed.

The **lateral head** of the deep digital flexor (Figs. 42, 43, 50, 53, 54) arises from the caudolateral border of the proximal two-thirds of the tibia, most of the proximal half of the fibula and the adjacent interosseous membrane. Its tendon begins as a wide expanse on the plantar side of the muscle but condenses distally. Medial to the tuber calcanei it is surrounded by the tarsal synovial sheath and bound in the groove over the sustentaculum tali of the calcaneus by the **flexor retinaculum.** At the level of the distal row of tarsal bones observe the tendon of the lateral head joining that of the medial head to form the deep digital flexor tendon. The courses, relations and attachments of the tendons distal to the tarsus are similar to those of the deep flexor tendon of the forelimb. Their dissection is not necessary.

The **medial head** of the deep digital flexor (Figs. 50, 54), smaller than the lateral head, lies between the lateral head and the popliteus. From the head of the fibula and the proximal end of the tibia it runs distomedially. Its tendon lies on the caudomedial side of the tibia. At the distal row of tarsal bones it unites with the tendon of the lateral head.

ORIGIN: The proximal two-thirds of the tibia, the proximal half of the fibula and the adjacent interosseous membrane.

INSERTION: The plantar surface of the base of each of the distal phalanges.

ACTION: To flex the digits and extend the tarsus.

4. The **popliteus** (Figs. 42-44, 46, 50, 52) is covered by the gastrocnemius and the superficial digital flexor and lies on the stifle joint capsule and the proximal tibia. It arises from the lateral condyle of the femur by a long tendon which should be isolated just cranial to the lateral ligament of the stifle. At the junction of the tendon with the muscle there is a **sesamoid** which articulates with the lateral condyle of the tibia. The popliteus inserts on the proximal third of the tibia.

ORIGIN: The lateral condyle of the femur.

INSERTION: The proximal third of the caudal surface of the tibia.

ACTION: To flex the stifle and rotate the leg.

JOINTS OF THE PELVIC LIMB

The ischium and pubis of the right and left sides are joined on the median plane at the **symphysis pelvis.**

The **sacroiliac joint** (Figs. 55, 56) is an articulation of stability rather than mobility. The right and left wings of the ilia articulate with the broad right and left wings of the sacrum. In the adult the apposed articular surfaces are united by fibrocartilage. Around the periphery of the articular areas, bands of strong collagenous tissue, the **dorsal** and **ventral sacroiliac ligaments,** reinforce the fibrocartilage. Do not dissect this joint now.

The **sacrotuberous ligament** (Fig. 55) runs from the transverse processes of the last sacral and first caudal vertebrae to the ischiatic tuberosity. It serves as an origin for several muscles which have been dissected.

The **hip** (Figs. 55, 56) is a ball-and-socket joint whose main movements are flexion and extension. The joint capsule passes from the neck of the femur to a line peripheral to the acetabular lip. Transect the deep gluteal and iliopsoas muscles at their insertions. Cut the hip joint capsule to expose the joint and associated ligaments.

The **ligament of the femoral head** (Fig. 56) is a strong strand of collagenous tissue which extends from the acetabular fossa to the fovea

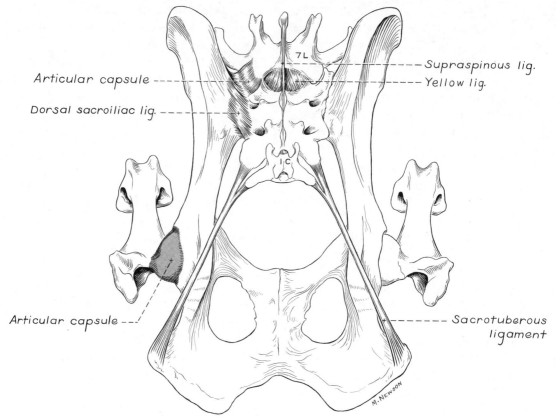

Figure 55.   Ligaments of pelvis, dorsal view.

capitis. At its acetabular attachment it may blend slightly with the transverse acetabular ligament. A synovial membrane covers it. Transect this ligament.

The **transverse acetabular ligament** (Fig. 56) is a small band which extends from one side of the acetabular notch to the opposite side. It continues the **acetabular lip** which deepens the acetabulum by forming a fibrocartilaginous border around it.

The **stifle** (Figs. 57-63) joint capsule forms three sacs. Two of these are between the femoral and tibial condyles and the third is beneath the patella. All three sacs communicate with each other. Between each femoral condyle and the corresponding tibial condyle there is a **meniscus** or **semilunar fibrocartilage.** These are C-shaped discs with thick peripheral margins and thin, concave central areas which compensate for the incongruence that exists between the femur and tibia.

Clean the remaining fascia and related muscle attachments away from the stifle. Transect the patellar ligament and reflect it proximally. Note the large quantity of fat between the patellar ligament and the joint capsule. Medial and lateral cartilaginous processes extend from the patella. Remove the fat around the joint, open the joint capsule and remove as much joint capsule as necessary to observe the following ligaments.

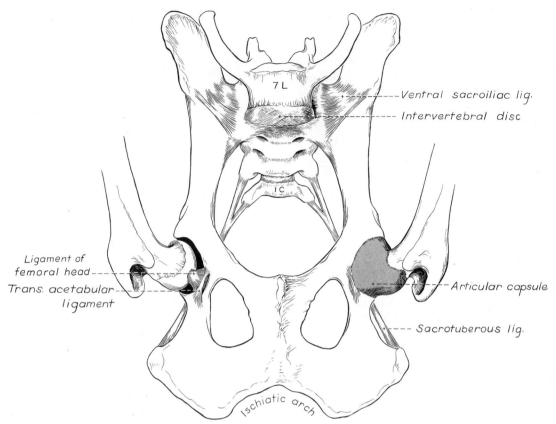

Figure 56. *Ligaments of pelvis, ventral view.*

Each meniscus attaches to the cranial and caudal intercondylar areas of the tibia. The caudal part of the lateral meniscus is attached to the intercondylar fossa of the femur by a **meniscofemoral ligament.**

The **femorotibial ligaments** are the collateral and cruciate ligaments. The **medial collateral ligament** fuses with the joint capsule and medial meniscus in its course from femur to tibia. The **lateral collateral ligament** attaches proximally to the femur and distally to the head of the fibula and the adjacent lateral condyle of the tibia. These may be transected to observe the deeper structures. The **cranial cruciate ligament** runs from the caudomedial part of the lateral femoral condyle diagonally across the intercondylar fossa of the femur to the cranial intercondylar area of the tibia. It crosses the **caudal cruciate ligament** which runs from the lateral surface of the medial femoral condyle caudodistally to the lateral edge of the popliteal notch of the tibia. The cruciates lie entirely within the joint capsule, where they are ensheathed by synovial membrane. Transect the cranial cruciate ligament and observe the increased craniocaudal mobility of the joint.

The two **tibiofibular joints** are a proximal joint between the head of the fibula and the lateral condyle of the tibia and a distal joint between the lateral malleolus of the fibula and the lateral surface of the distal end

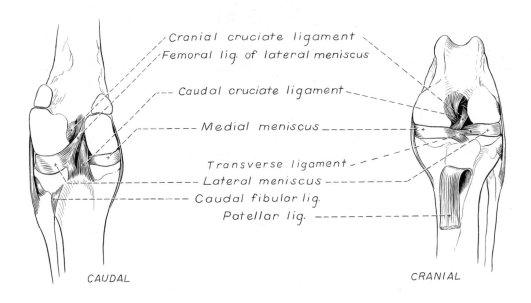

Cranial cruciate ligament
Femoral lig. of lateral meniscus
Caudal cruciate ligament
Medial meniscus
Transverse ligament
Lateral meniscus
Caudal fibular lig.
Patellar lig.

CAUDAL   CRANIAL

Figure 57.   Ligaments of stifle, caudal view.   *Figure 58.*   Ligaments of stifle, cranial view.

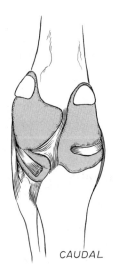

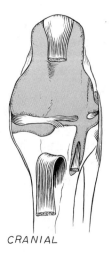

CAUDAL   CRANIAL

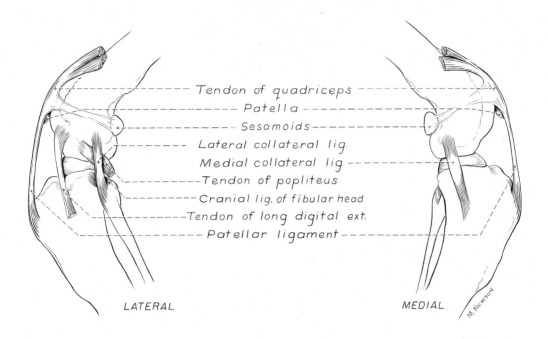

Tendon of quadriceps
Patella
Sesamoids
Lateral collateral lig.
Medial collateral lig.
Tendon of popliteus
Cranial lig. of fibular head
Tendon of long digital ext.
Patellar ligament

LATERAL

MEDIAL

Figure 59. *Ligaments of stifle, lateral view.*   **Figure 60.** *Ligaments of stifle, medial view.*

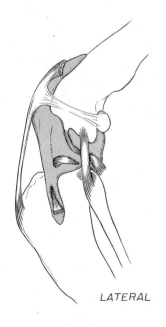

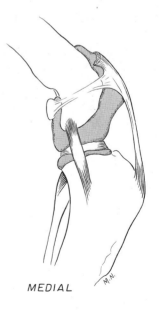

LATERAL

MEDIAL

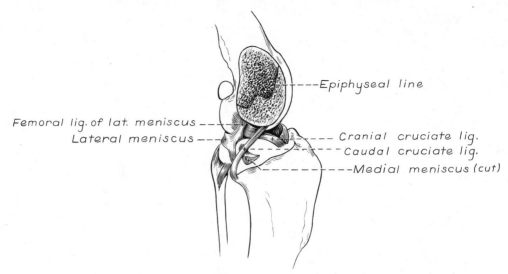

Figure 61. *Cruciate and meniscal ligaments, left stifle.*

of the tibia. Throughout the length of the interosseous space between the tibia and fibula is a sheet of fibrous tissue uniting the two bones, the **interosseous membrane** of the crus.

In the **tarsal joint** or hock the joint capsule is largest between the tibial cochlea and the trochlea of the talus. Greatest movement takes place here.

BONES OF THE VERTEBRAL COLUMN

The **vertebral column** (see Fig. 2) consists of approximately 50 irregular bones. The vertebrae are arranged in five groups: **cervical, thoracic, lumbar, sacral** and **caudal.** The first letter or abbreviation of the word designating each group followed by the number of vertebrae in each group expresses a vertebral formula. That of the dog is $C_7 \, T_{13} \, L_7 \, S_3 \, Cd_{20}$. The number 20 for the caudal vertebrae is arbitrary; many dogs have fewer. The three sacral vertebrae fuse to form what is considered a single bone, the **sacrum.**

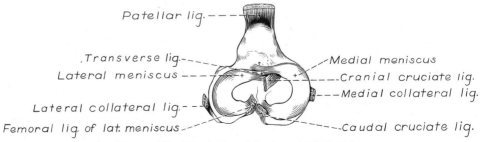

Figure 62. *Menisci and ligaments, left tibia.*

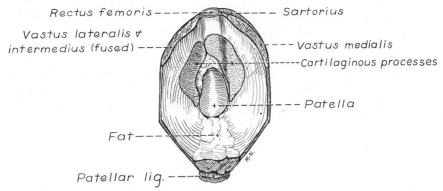

Rectus femoris ---- --- Sartorius

Vastus lateralis &
intermedius (fused) -- --- Vastus medialis

----- Cartilaginous processes

--- Patella

Fat ---

Patellar lig. ---

Figure 63. *Left patella, caudal view.*

A typical vertebra consists of a body or centrum, a vertebral arch consisting of right and left pedicles and laminae, and transverse, spinous and articular processes.

The **body** of a vertebra is constricted centrally. The cranial extremity is convex and the caudal extremity is concave. Adjacent vertebrae are connected by intervertebral disks, each of which is a fibrocartilaginous structure composed of a soft center, the **nucleus pulposus,** surrounded by concentric layers of tough fibrous tissue, the **annulus fibrosus.** These intervertebral disks will be dissected when the spinal cord is studied.

The **vertebral arch** is subdivided into basal parts, the **pedicles,** and the dorsal portion, formed by two **laminae.** Together with the body the vertebral arch forms a short tube, the **vertebral foramen.** All the vertebral foramina join to form the **vertebral canal.** The pedicles of each vertebra extend from the dorsolateral surface of the body. They present smooth-surfaced notches. The **cranial vertebral notches** are shallow; the **caudal vertebral notches** are deep. When the vertebral column is articulated the notches of adjacent vertebrae and the intervening fibrocartilage form the right and left **intervertebral foramina.** Through these pass the spinal nerves and blood vessels. The dorsal part of the vertebral arch is composed of right and left laminae which unite to form a spinous process. Each typical vertebra has, in addition to the dorsally located **spinous process** or spine, paired **transverse processes** which project laterally from the region where the arch joins the vertebral body. Farther dorsally on the arch, at the junction of pedicle and lamina, are located the **articular processes.** There are two of these on each side of the vertebra—a cranial pair whose articulating facets point dorsally or medially and a caudal pair whose facets are directed ventrally or laterally.

CERVICAL VERTEBRAE

There are seven cervical vertebrae in domestic animals. The **atlas** (Fig. 64), or first cervical vertebra, is atypical in both structure and function. It articulates with the skull cranially. Its chief peculiarities are modified articular processes, lack of a spinous process and reduction of its body. The lateral parts are thick, forming the **lateral masses.** These are

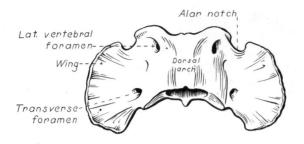

Figure 64. *Atlas, dorsal view.*

united by the **dorsal** and **ventral arches.** The shelf-like transverse processes or wings project from the lateral masses. The two **cranial articular foveae** articulate with the occipital condyles of the skull to form the atlanto-occipital joint, of which the main movement is flexion and extension. The **caudal articular foveae** consist of two shallow glenoid cavities which form a freely movable articulation with the second cervical vertebra. Rotatory movement occurs at this joint. Examine the caudal part of the dorsal surface of the ventral arch for the **fovea dentis.** This is concave from side to side and articulates with the dens of the second cervical vertebra. This articular area blends with those of the caudal surfaces of the lateral masses. Besides the large vertebral foramen through which the spinal cord passes, there are two pairs of foramina in the atlas. The **transverse foramina** are actually short canals which are located just lateral to the lateral masses and which pass obliquely through the transverse processes of the atlas. The **lateral vertebral foramina** perforate the cranial part of the dorsal arch. The first cervical spinal nerves pass through these foramina.

The **axis,** or second cervical vertebra (Fig. 65), presents an elongated, ridge-like spinous process as its most prominent characteristic. It is also peculiar in that cranially the body projects forward in a peg-like eminence, the **dens.** The ventral surface of the dens is articular, while the tip and dorsal surface may be rough because of ligamentous attachment. The **cranial articular surface** is located on the body and is continuous with the articular area of the dens. The caudal part of the vertebral arch has two articular processes which face ventrolaterally. At the root of the transverse process is the small transverse foramen. The cranial vertebral notch concurs with that of the atlas to form the intervertebral foramen for

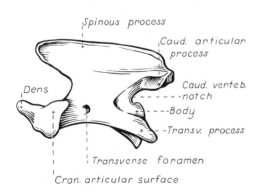

Figure 65. *Axis, lateral view.*

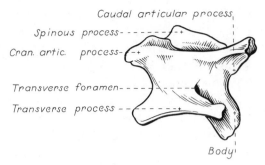

Figure 66. *Fourth cervical vertebra, lateral view.*

the transmission of the second cervical nerve. The caudal notch concurs with that of the third cervical vertebra to form the third intervertebral foramen, through which the third cervical nerve passes. Other features of this bone are similar to those of a typical vertebra.

The middle three cervical vertebrae (Fig. 66) differ little from a typical vertebra. The spinous processes are low but gradually increase in height from third to fifth. The transverse processes are two-pronged and are perforated at the base by a transverse foramen.

The sixth cervical vertebra has a high spine and an expanded **ventral lamina** of the transverse process. The seventh cervical vertebra lacks transverse foramina and has the highest cervical spine.

The **cranial articular processes** of cervical vertebrae 3 to 7 face dorsally and cranially in apposition to the **caudal articular processes** of adjacent vertebrae. There is a tubercle on the caudal midventral aspect of the vertebral body.

THORACIC VERTEBRAE

There are 13 thoracic vertebrae (Fig. 67); the first nine are similar. The bodies of the thoracic vertebrae are short. The first through the tenth have a **cranial** and **caudal costal fovea** for rib articulation. Since the foveae of two adjacent vertebrae form the articular surface for the head of one rib, they are sometimes called demifacets. The body of the eleventh vertebra frequently lacks caudal foveae, while the twelfth and thirteenth

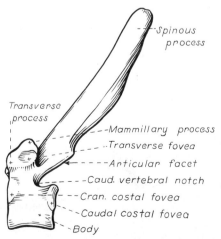

Figure 67. *Sixth thoracic vertebra, lateral view.*

have only one complete fovea on each side. The head of the first rib articulates between the last cervical and first thoracic vertebrae. The tubercles of the ribs articulate with the transverse processes of the thoracic vertebrae of the same number in all instances.

The **spine** is the most conspicuous feature of each of the first nine thoracic vertebrae. The massiveness of the spines gradually decreases with successive vertebrae, but there is little change in length and direction until the seventh or eighth is reached. The spines then become progressively shorter and incline caudally through the ninth and tenth. The spine of the eleventh thoracic vertebra is nearly perpendicular to the long axis of that bone. This vertebra is designated the **anticlinal vertebra.** All spines caudal to the eleventh point cranially; all spines cranial to the eleventh point caudally.

The transverse processes are short, blunt and irregular. All contain costal foveae for articulation with the tubercles of the ribs.

The **articular processes** are located at the junctions of the pedicles and the laminae. The cranial pair are nearly confluent at the median plane on thoracic vertebrae 3 to 10. Since the caudal articular processes articulate with the cranial ones of the vertebra behind, they are similar in shape but face in the opposite direction. The joints between thoracic vertebrae 10 and 13 are conspicuously modified since the facets of the caudal articular processes are located on the lateral surfaces. This type of interlocking articulation allows flexion and extension of the caudal-thoracic and lumbar region while limiting lateral movement.

LUMBAR VERTEBRAE

The lumbar vertebrae (Fig. 68) have longer bodies than thoracic vertebrae. The transverse processes are directed cranially as well as ventrolaterally. The articular processes lie mainly in sagittal planes. The caudal processes protrude between the cranial ones of succeeding vertebrae.

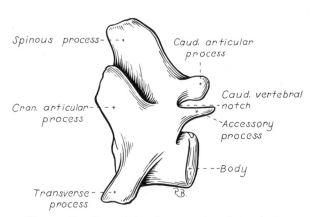

Figure 68. Fourth lumbar vertebra, lateral view.

SACRUM

The sacrum (Figs. 69, 70) results from the fusion of the bodies and processes of three vertebrae. This bone lies between the ilia and firmly articulates with them. The body of the first segment is larger than the combined bodies of the other two segments. The three are united to form a concave ventral surface.

The dorsal face presents several markings which result from the fusion of the three sacral vertebrae. The **median sacral crest** represents the fusion of the three spinous processes. The dorsal surface also bears two pairs of **dorsal sacral foramina,** which transmit the dorsal branches of the sacral spinal nerves.

The pelvic face has two pairs of **pelvic sacral foramina.** They transmit the ventral branches of the first two sacral spinal nerves. The wing of the sacrum is the enlarged lateral part which bears a large, rough facet, the **auricular face,** which articulates with the ilium.

The **base** of the sacrum faces cranially. The ventral part of the base has a transverse ridge, the **promontory.** This, with the ilia, forms the dorsal boundary of the pelvic inlet.

CAUDAL VERTEBRAE

The average number of caudal vertebrae (Fig. 71) in the dog is 20. These lose their distinctive features as one proceeds caudally.

RIBS

All 13 pairs of ribs (Fig. 72) have dorsal bony and ventral cartilaginous parts. The cartilaginous parts are called **costal cartilages.** The first nine pairs of ribs articulate directly with the sternum. The costal carti-

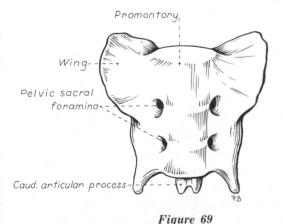

Figure 69

Sacrum, ventral view.

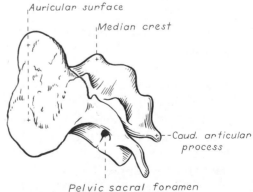

Figure 70

Sacrum, lateral view.

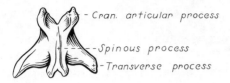

Cran. articular process

Spinous process

Transverse process

Figure 71. Third caudal vertebra, dorsal view.

lages of the tenth, eleventh and twelfth unite with one another to form the **costal arch.** The thirteenth rib often ends freely in the flank.

The bony part of a typical rib presents a head, neck, tuberculum and body.

The **head** articulates with the costal foveae of two contiguous vertebrae and the intervening fibrocartilage. The **tuberculum** of the rib articulates with the costal fovea of the transverse process of the vertebra of the same number. Between the head and tuberculum of the rib is the **neck**.

STERNUM

The sternum is composed of eight unpaired segments, the **sternebrae** (Fig. 72). Consecutive sternebrae are joined by the **intersternebral cartilages.** The first sternebra, also known as the **manubrium,** ends cranially in a club-like enlargement. The last sternebra is flattened dorsoventrally and is called the **xiphoid process.** The caudal end of this process is continued by a thin plate of cartilage.

MUSCLES OF THE TRUNK

The muscles of the trunk, or axial muscles, are divided in a morphological sense into hypaxial and epaxial groups. The epaxial muscles lie dorsal to the transverse processes of the vertebrae and function mainly as extensors of the vertebral column. The hypaxial group embraces all other trunk muscles not included in the epaxial division. These muscles are located ventral to the transverse processes and include those of the abdominal and thoracic walls.

The **superficial fascia of the trunk** covers the thorax and abdomen subcutaneously. It is continuous cranially and caudally with the superficial fasciae of the thoracic limb, the neck and the pelvic limb. Clean this fascia from the thorax and abdomen. In doing so detach the origin of the latissimus dorsi from the last few ribs and reflect it to the middorsal line where it arises from the lumbar vertebral spines via the superficial leaf of the thoracolumbar fascia.

The **deep fascia of the trunk** (Fig. 38), the thoracolumbar fascia, is attached to the ends of the spinous and transverse processes of the thoracic and lumbar vertebrae. It passes over the epaxial musculature to the lateral thoracic and abdominal wall where it serves as origin for several muscles.

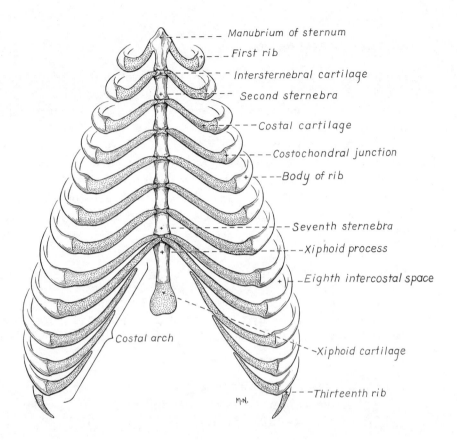

Manubrium of sternum
First rib
Intersternebral cartilage
Second sternebra
Costal cartilage
Costochondral junction
Body of rib
Seventh sternebra
Xiphoid process
Eighth intercostal space
Xiphoid cartilage
Thirteenth rib
Costal arch

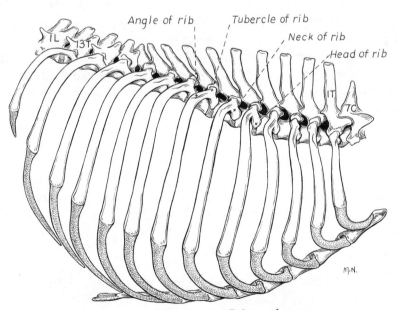

Angle of rib   Tubercle of rib
Neck of rib
Head of rib

Figure 72. Ribs and sternum.

HYPAXIAL MUSCLES

MUSCLES OF THE NECK

The **longus capitis** (Fig. 77) lies on the lateral and ventral surfaces of cervical vertebrae, lateral to the longus colli. It arises from the transverse processes of the cervical vertebrae and inserts on the muscular tubercle of the basioccipital bone.

The **longus colli** covers the ventral surfaces of the vertebral bodies from the sixth thoracic vertebra cranially to the atlas. It consists of many overlapping fascicles that attach to the vertebral bodies or the transverse processes. The most cranial cervical bundles attach to the atlas. Expose this muscle by reflecting the trachea and the esophagus and related soft tissues.

MUSCLES OF THE THORACIC WALL

1. The **scalenus** (Figs. 73, 77) lies ventral to the origin of the serratus ventral. It attaches to the first few ribs and the transverse processes of the cervical vertebrae and is divided into three slips. It is a muscle of inspiration.

2. The **serratus ventralis** (Fig. 73) is a large, fan-shaped muscle with an extensive origin on the neck and trunk. Its insertion on the serrated face of the scapula has been observed (see Fig. 14).

3. The **serratus dorsalis** arises by a broad aponeurosis from the tendinous raphe of the neck and from the thoracic spines and inserts on the proximal portions of the ribs. It consists of two portions:

The **serratus dorsalis cranialis** (Fig. 73) lies on the dorsal surface of the cranial thorax. It arises by a broad aponeurosis from the thoracolumbar fascia. It runs caudoventrally and inserts by distinct serrations on the craniolateral surfaces of the ribs. It lifts the ribs for inspiration. Transect this muscle at the beginning of its muscle fibers and reflect both portions.

The **serratus dorsalis caudalis** is found on the dorsal surface of the caudal thorax and consists of distinct muscle leaves which arise by an aponeurosis from the thoracolumbar fascia, course cranioventrally and insert on the caudal borders of the last three ribs. It functions in drawing the last three ribs caudally in expiration.

4. There are 12 **external intercostal muscles** (Fig. 73) on each side of the thoracic wall. Their fibers run caudoventrally from the caudal border of one rib to the cranial border of the rib behind. Ventrally they end near the costochondral junction. They function in respiration by drawing the ribs together, and their overall effect depends upon the fixation of the rib cage.

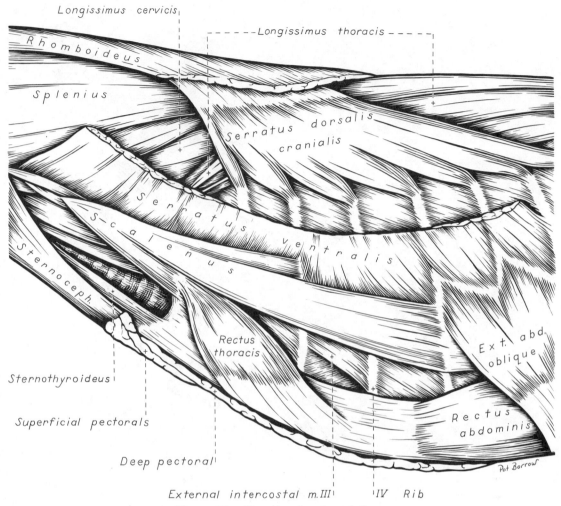

Figure 73. Muscles of neck and thorax.

5. The **internal intercostal muscles** (see Fig. 82) are easily differentiated from the external intercostal muscles because their fibers run ventrocranially. Medial to most of the internal intercostal muscles is the pleura. It attaches to the muscles and ribs by the endothoracic fascia. The internal intercostal muscles extend the whole distance of the intercostal spaces. These muscles function in a manner similar to that of the external intercostal muscles.

Expose the fifth external intercostal muscle and reflect it to observe the internal intercostal.

MUSCLES OF THE ABDOMINAL WALL

Four thin, muscular sheets enclose the abdomen. The four abdominal muscles, named from without inward, are the external abdominal oblique, the internal abdominal oblique, the rectus abdominis and the transversus

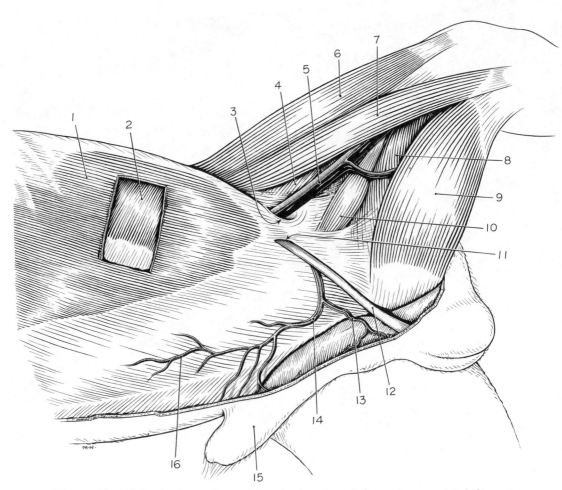

Figure 74. *Abdominal muscles and inguinal region of the male, superficial dissection.*

1. *External abdominal oblique*
2. *Internal abdominal oblique*
3. *Vascular lacuna*
4. *Vastus medialis*
5. *Femoral artery and vein in femoral triangle*
6. *Cranial part of sartorius*
7. *Caudal part of sartorius*
8. *Adductor*
9. *Gracilis*
10. *Pectineus*
11. *Superficial inguinal ring*
12. *Parietal vaginal tunic*
13. *Cranial scrotal artery and vein*
14. *External pudendal artery and vein, genital branch*
15. *Prepuce*
16. *Caudal superficial epigastric artery and vein*

abdominis. When they contract they aid in urination, defecation, parturition, respiration and locomotion. These muscles are covered superficially by the abdominal fasciae and deeply by the transverse fascia.

1. The **external abdominal oblique** (Figs. 73-75) covers the ventral half of the lateral thoracic wall and the lateral part of the abdominal wall. The costal part arises from the last ribs. The lumbar part arises from the last rib and from the thoracolumbar fascia. The fibers of this muscle run caudoventrally. In the ventral abdominal wall it forms a wide aponeurosis which inserts on the linea alba and the cranial pubic ligament. The **linea alba** is the midventral aponeurosis of the abdominal muscles which extends from the xiphoid process to the symphysis pelvis. The aponeurosis of the external abdominal oblique, combined with that of the internal abdominal oblique, forms most of the external lamina of the sheath of the rectus abdominis.

Caudoventrally, just cranial to the iliopubic eminence and lateral to the midline, the aponeurosis of the external abdominal oblique separates into its two parts, which then come together to form the **superficial inguinal ring** (Figs. 74, 75). This is the external opening of a very short, natural passageway through the abdominal wall, the **inguinal canal.** The internal opening and the boundaries of the canal will be seen when the abdominal cavity is opened. A blind extension of peritoneum protrudes through the inguinal canal to a subcutaneous position outside the body wall. This becomes the **vaginal tunic** (Fig. 75) in the male and the **vaginal process** in the female. In the male it is accompanied by the testis and spermatic cord which it envelops. In the female, the vaginal tunic envelops the round ligament of the uterus and a varying amount of fat and ends blindly a short distance from the vulva. In both sexes the external pudendal artery and vein and the genital nerve also pass through the inguinal canal (Figs. 74, 117).

Clean the surface of the aponeurosis of the external abdominal oblique, and identify the superficial inguinal ring and vaginal tunic. The vaginal tunic is covered by spermatic fascia which is continuous with the abdominal fascia.

Transect the external abdominal oblique close to its origin. Reflect this muscle from the internal abdominal oblique to the line of fusion between their aponeuroses. The fused aponeuroses cover the rectus abdominis.

The **inguinal ligament** consists of fibers running from the tuber coxae to the iliopubic eminence. The caudal border of the external abdominal oblique may insert on the inguinal ligament. Proximally, the internal abdominal oblique muscle arises from this ligament. Distally, the ligament is interposed between the **vascular lacuna** (Fig. 74), the base of the femoral triangle containing the femoral vessels that run to and from the hind limb, and the **superficial inguinal ring.** It thus forms part of the cranial border of the vascular lacuna and the caudal border of the inguinal canal. Transect the aponeurosis of insertion 2 cm. cranial to the inguinal ligament and the superficial inguinal ring. Observe these structures.

2. The **internal abdominal oblique** (Figs. 74, 75) arises from the thoracolumbar fascia caudal to the last rib in common with the lumbar

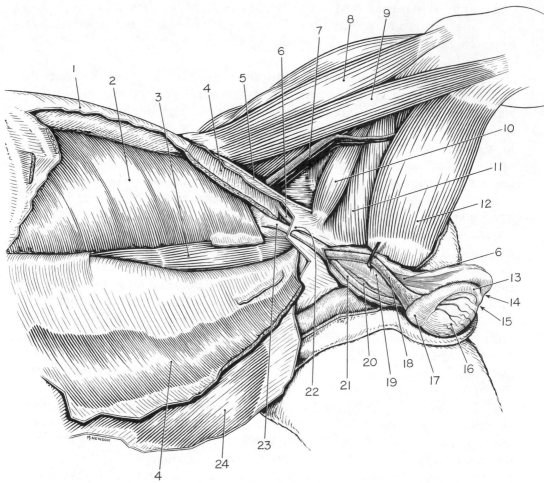

Figure 75. *Abdominal muscles and inguinal region of the male, deep dissection.*

1. Thoracolumbar fascia
2. Transversus abdominis
3. Rectus abdominis
4. Internal abdominal oblique (transected and reflected)
5. Inguinal ligament
6. Cremaster muscle
7. Femoral artery and vein
8. Cranial part of sartorius
9. Caudal part of sartorius
10. Pectineus
11. Adductor
12. Gracilis
13. Tail of epididymis
14. Caudal ligament of epididymis
15. Proper ligament of testis
16. Testis in visceral vaginal tunic
17. Head of epididymis
18. Testicular artery and vein in visceral vaginal tunic (mesorchium)
19. Mesorchium
20. Mesoductus deferens
21. Ductus deferens in visceral vaginal tunic
22. Superficial inguinal ring
23. Parietal vaginal tunic
24. External abdominal oblique (reflected)

part of the external abdominal oblique, and from the tuber coxae and adjacent portion of the inguinal ligament. Its fibers run cranioventrally. It inserts by a wide aponeurosis on the costal arch, on the rectus abdominis and on the linea alba, in common with that of the external abdominal oblique to which it is fused.

Transect the muscle 2 cm. from its origin and reflect it ventrally to the rectus abdominis. In doing so detach its insertions on the ribs. Separate it from all underlying structures except the rectus abdominis. Study the aponeurosis of the internal abdominal oblique which contributes to the external lamina of the sheath of the rectus abdominis. Note how the caudal border of the muscle forms the cranial border of the inguinal canal.

3. The **transversus abdominis** (Figs. 38, 75) is medial to the internal abdominal oblique and the rectus abdominis. Its fibers run transversely. The muscle arises dorsally from the medial surfaces of the last four or five ribs and from the transverse processes of all the lumbar vertebrae by means of the thoracolumbar fascia. Its aponeurosis attaches to the linea alba after crossing the internal surface of the rectus abdominis.

4. The **rectus abdominis** (Figs. 73, 75) extends from the pubis to the sternum and flexes the thoracolumbar part of the vertebral column. Observe its cranial aponeurosis from the first few ribs and sternum. The rectus abdominis has distinct transverse tendinous intersections. In the umbilical region the external lamina of the sheath of the rectus abdominis is formed by the fused aponeuroses of the oblique muscles. The internal lamina is formed by the aponeurosis of the transversus abdominis.

The Inguinal Canal. The inguinal canal is a slit between the abdominal muscles. It extends from the deep to the superficial inguinal ring. The **superficial inguinal ring** in the aponeurosis of the external abdominal oblique has already been dissected. The **deep inguinal ring** is formed on the inside of the abdominal wall by the annular reflection of transversalis fascia onto the vaginal tunic. This fascia lies between the transversus abdominis and peritoneum. The ring is a boundary and not a distinct anatomical structure. The inguinal canal is bounded laterally by the aponeurosis of the external abdominal oblique, cranially by the internal abdominal oblique as it arises from the inguinal ligament, caudally by the inguinal ligament and medially by the lateral border of the rectus abdominis and by the transversalis fascia and peritoneum. The vaginal tunic and the **spermatic cord** pass obliquely caudoventrally through the inguinal canal.

EPAXIAL MUSCLES

The dorsal trunk musculature associated with the vertebral column and ribs may be divided into three longitudinal muscle masses on each side. Each is comprised of many overlapping fascicles. These three columns include the lateral **iliocostalis** system, the intermediate **longissimus** system and the medial **transversospinalis** system (Fig. 76). Various fusions occur between these columns, giving rise to different muscle patterns. These muscles act as extensors of the vertebral column and also produce lateral movements of the trunk when contracting on only one side.

ILIOCOSTALIS SYSTEM

1. The **iliocostalis lumborum** (Fig. 76) arises from the wing of the ilium in common with the longissimus lumborum and inserts on the transverse processes of the lumbar vertebrae and the last four or five ribs. In the lumbar region this muscle is fused with the longissimus lumborum. The thoracolumbar fascia covers these muscles. Remove this fascia and any underlying fat to expose the glistening aponeurosis of these fused muscles which attaches to the crest of the ilium, and the spines of the lumbar and last four or five thoracic vertebrae. The cranial end of this muscle is distinctly separated from the longissimus lumborum as it inserts on the ribs. Expose this insertion.

2. The **iliocostalis thoracis** (Fig. 77) is a long, narrow muscle mass extending from the twelfth rib to the transverse process of the seventh cervical vertebra. Individual components of the muscle extend between and overlap the ribs. Identify the boundaries of this muscle mass.

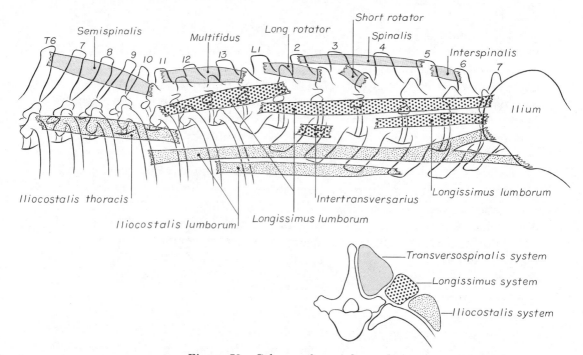

Figure 76. Schema of epaxial muscles.

LONGISSIMUS SYSTEM

The longissimus is the intermediate portion of the epaxial musculature. Lying medial to the iliocostalis, its overlapping fascicles extend from the ilium to the head. It consists of three major regional divisions: **thoracolumbar, cervical** and **capital.**

1. The **longissimus thoracis et lumborum** (Fig. 77) arises from the crest and medial surface of the wing of the ilium and, by means of an aponeurosis, from the supraspinous ligament and the spines of the lumbar and thoracic vertebrae. Its fibers course craniolaterally. Superficially,

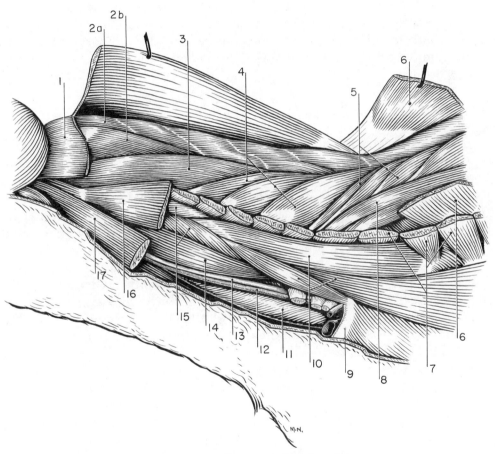

Figure 77. *Deep muscles of neck.*

1. *Splenius*
2. *Semispinalis capitis:*
   a. *Biventer cervicis*
   b. *Complexus*
3. *Longissimus capitis*
4. *Longissimus cervicis*
5. *Longissimus thoracis*
6. *Serratus dorsalis cranialis*
7. *Serratus ventralis*
8. *Iliocostalis thoracis*
9. *First rib*
10. *Scalenus*
11. *Esophagus*
12. *Common carotid a.*
13. *Vagosympathetic trunk*
14. *Longus capitis*
15. *Intertransversarius*
16. *Omotransversarius*
17. *Cleidomastoideus*

only a shallow furrow is seen to separate the longissimus and iliocostalis muscles in the lumbar region. This division of the longissimus inserts on various processes of the lumbar and thoracic vertebrae. The thoracic portion may be seen inserting on the ribs just medial to the iliocostalis thoracis.

2. The **longissimus cervicis** (Fig. 77), the cranial continuation of the longissimus muscle into the neck, consists of four fascicles so arranged that the caudal fascicles partly cover those which lie directly cranioventral to them. They lie in the angle between the cervical and thoracic vertebrae and insert on the transverse processes of the last few cervical vertebrae.

3. The **longissimus capitis** (Fig. 77) is a distinct muscle medial to the longissimus cervicis and splenius muscles. It extends from the first three thoracic vertebrae to the mastoid part of the temporal bone. It is firmly united with the splenius as it passes over the wing of the atlas to its insertion.

TRANSVERSOSPINALIS SYSTEM

The transversospinalis system, the most medial and deep epaxial muscle mass, consists of a number of different groups of muscles which join one vertebra with another or span one or more vertebrae. This complex system extends from the sacrum to the head. Included are muscles whose names depict their attachments or the functions of their fascicles: spinalis, semispinalis, multifidus, rotatores, interspinalis and intertransversarius. Only a few of these muscles will be dissected.

The **splenius** (Fig. 73) is a rather large muscle on the dorsolateral surface of the neck, deep to the rhomboideus capitis and the serratus dorsalis cranialis. Its fibers extend in a slightly cranioventral direction from the third thoracic vertebra to the skull. The muscle arises from the cranial border of the thoracolumbar fascia, the spines of the first three thoracic vertebrae and the entire median raphe of the neck. It inserts on the nuchal crest and mastoid part of the temporal bone. Transect the splenius 2 cm. caudal to its insertion and reflect the muscle mass to the midline.

The **semispinalis capitis** (Fig. 77) is a member of the cervical portion of the transversospinalis group. It lies deep to the splenius and extends from the thoracic vertebrae to the head. It is divided into the biventer cervicis and the complexus. These can be separated to their insertions despite their intimate connections to each other.

The **biventer cervicis** is dorsal to the complexus and has tendinous intersections. It arises from thoracic vertebrae and inserts on the caudal surface of the skull. Transect the muscle and reflect it.

The **complexus** is ventral to the biventer and arises from cervical vertebrae. It inserts on the nuchal crest. Transect this muscle and reflect it.

The **nuchal ligament** may be seen extending from the tip of the spinous process of the first thoracic vertebra to the broad caudal end of the spine of the axis. It is a laterally compressed, paired, yellow elastic band lying between the medial surfaces of the two semispinalis capitis muscles.

The **supraspinous ligament** continues the nuchal ligament caudally and extends from the spinous process of the first thoracic vertebra to the caudal vertebrae. It passes from one spinous process to another. It will be seen later in the preparation for the spinal cord dissection.

JOINTS OF THE VERTEBRAL COLUMN

These will be studied with the dissection of the spinal cord.

THE NECK, THORAX AND THORACIC LIMB

Make a skin incision from the midventral line at the level of the thoracic limb to the medial side of the right elbow joint. Make a circular skin incision at the elbow joint. Make transverse skin incisions from midventral to middorsal lines at the level of the umbilicus and at the cranial part of the neck. Reflect the skin flap to the middorsal line leaving the cutaneous muscles and superficial fascia on the dog.

In the following dissection of vessels and nerves the arteries can be recognized by the red latex which was injected into the arterial system. The veins will sometimes contain dark-colored clotted blood.

VESSELS AND NERVES OF THE NECK

There are eight pairs of cervical spinal nerves in the dog. The first cervical nerve passes through the lateral vertebral foramen of the atlas. The remaining nerves pass through succeeding intervertebral foramina. The eighth cervical nerve emerges from the intervertebral foramen between the seventh cervical and first thoracic vertebrae. Immediately upon leaving the foramina the nerves divide into large ventral and small dorsal branches. The dorsal branches supply structures dorsal to the vertebrae (Fig. 82).

Uncover the ventral branch of the **second cervical nerve** (Figs. 78, 79). This lies along or deep to the middle of the caudoventral border of the platysma, which is dorsal to the external jugular vein. Separate the overlying fascia until the nerve is found. It emerges between the cleido-mastoideus and the omotransversarius. The ventral branch of the second cervical nerve divides into two cutaneous branches: (1) The **great auricular nerve** extends toward the ear. It branches and supplies the skin of the neck, the ear and the back of the head. Trace the nerve as far as present

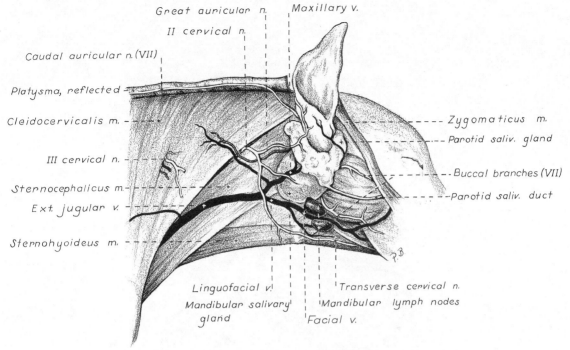

Great auricular n.   *Maxillary v.*

II cervical n.

Caudal auricular n. (VII)

Platysma, reflected

Cleidocervicalis m.

III cervical n.

Sternocephalicus m.

Ext. jugular v.

Sternohyoideus m.

Zygomaticus m.

Parotid saliv. gland

Buccal branches (VII)

Parotid saliv. duct

Linguofacial v.

Mandibular salivary gland

Transverse cervical n.

Mandibular lymph nodes

Facial v.

Figure 78. *Superficial structures of neck.*

muscle and skin reflections will allow. (2) The **transverse cervical nerve** branches to the skin of the cranioventral part of the neck.

The **external jugular vein** (Figs. 78, 80), on the side of the neck, is formed by the **linguofacial** and **maxillary** veins. The ovoid body which lies in the fork formed by these veins is the **mandibular salivary gland** (Fig. 78). The **mandibular lymph nodes** (Fig. 78) lie on both sides of the linguofacial vein.

Ligate and transect the external jugular vein at its approximate middle and reflect each end. Notice the **cephalic vein** (Fig. 80) which enters the jugular after crossing the shoulder from its course up the arm. It may be transected and reflected. Free the sternocephalicus and transect it 3 cm. from its origin. Carefully turn it craniodorsally to a point cranial to the place where the second cervical nerve crosses the muscle. Transect the brachiocephalicus 1 cm. cranial to the clavicular intersection.

The **superficial cervical lymph nodes** lie in the areolar tissue cranial to the shoulder. They lie deep to the cervical parts of the brachiocephalic and omotransverse muscles and receive lymph drainage from the cutaneous area of the scapular region and neck.

The **accessory** or **eleventh cranial nerve** (Fig. 79) is a large nerve found deep to the cranial part of the sternocephalicus. As it emerges from the cleidomastoideus it crosses the second cervical nerve, runs along the dorsal border of the omotransversarius and terminates in the thoracic part of the trapezius. The accessory nerve is the only motor nerve to the trapezius and, in addition, supplies in part the omotransversarius, the

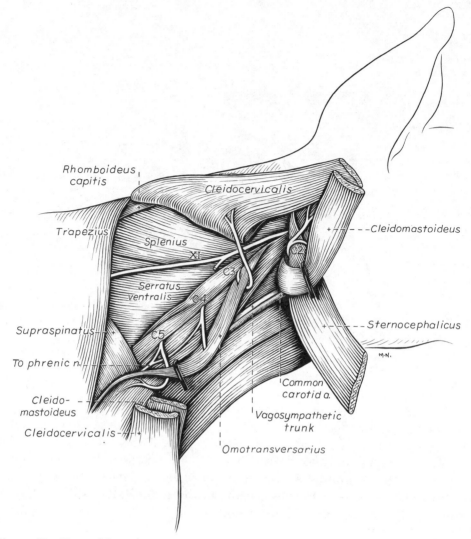

Figure 79. *Ventral branches of cervical spinal nerves emerging from lateral neck muscles.*

mastoid and cervical part of the brachiocephalicus and the sternocephalicus.

Free the ventral border of the omotransversarius and lift it. Observe the ventral branches of the **third, fourth** and **fifth cervical nerves,** which are distributed in a segmental manner to the muscles and skin of the neck. The third and fourth nerves, after emerging from the intervertebral foramina, pass through the deep fascia and the omotransversarius.

Transect the fused sternohyoideus and sternothyroideus 2 cm. from their origin and reflect them to their insertions. Parts of the trachea, larynx, thyroid gland, esophagus and carotid sheath are exposed. Identify these structures on your specimen. Note the common carotid artery dorsal to the sternothyroideus. Bound to its medial side is the **vagosympathetic nerve trunk.** The **medial retropharyngeal lymph node** lies opposite the larynx, ventrolateral to the carotid sheath.

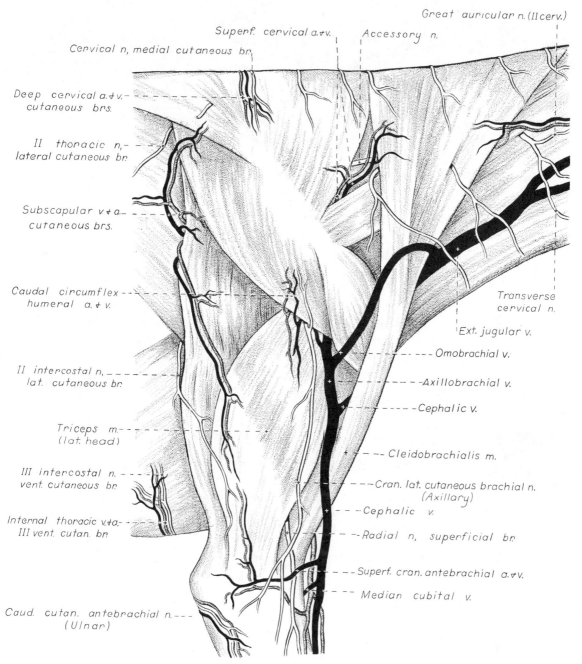

Figure 80. *Superficial structures of scapula and arm, lateral view.*

THE THORAX

SUPERFICIAL VESSELS AND NERVES OF THE THORACIC WALL

Before dissecting the thoracic nerves and vessels study Figures 81 to 84, which show the pattern of distribution of these structures. Notice that the artery and nerve of each intercostal space divide into dorsal and

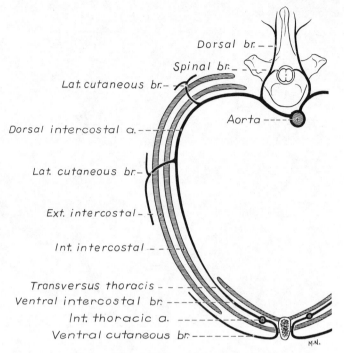

Figure 81. *Arteries of thoracic wall.*

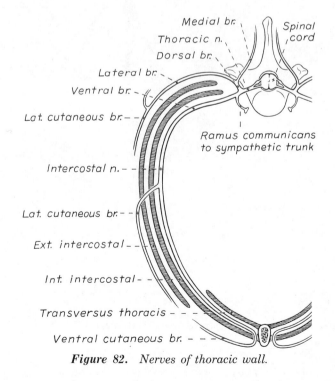

Figure 82. *Nerves of thoracic wall.*

ventral branches. The dorsal branches end in the epaxial muscles and the vertebrae. The ventral branches descend in the intercostal spaces. The dorsal and ventral arterial branches are derived from the dorsal intercostal arteries. The first three dorsal intercostal arteries come from a branch of the costocervical trunk, the remaining nine come from the aorta. The dorsal intercostal arteries and veins have lateral cutaneous branches, and ventrally they anastomose with ventral intercostal branches from the internal thoracic artery and vein. The ventral cutaneous branches with mammary radicles are branches of the internal thoracic vessels.

The dorsal and ventral nerve branches are derived from the spinal nerve as it emerges from the intervertebral foramen. The ventral branches of the first 12 thoracic nerves are intercostal nerves and have lateral and ventral cutaneous branches and branches medial to these which go largely to muscles.

The **cranial thoracic mamma** is supplied by the fourth, fifth and sixth ventral and lateral cutaneous vessels and nerves, and branches of the lateral thoracic vessels. The latter are from the axillary vessels, which will be dissected later.

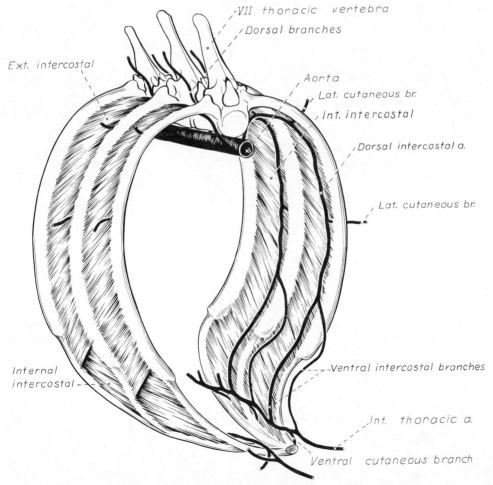

Figure 83. *Schema of intercostal structures.*

The **caudal thoracic mamma** is supplied in a similar manner from the sixth and seventh cutaneous nerves and vessels. In addition, mammary branches of the cranial superficial epigastric vessels supply this mamma. The parent vessels will be dissected later.

The **axilla** is the space between the thoracic limb and the thoracic wall. It is bounded ventrally by the pectoral muscles and dorsally by the attachment of the serratus ventralis to the scapula. Cranially it extends under the muscles which extend from the arm to the neck. Caudally, a similar extension is found under the latissimus dorsi and cutaneus trunci.

Dorsal and lateral rows of lateral cutaneous branches of intercostal nerves, arteries and veins emerge at regular intervals between the ribs and supply the cutaneous muscle, subcutaneous tissue and skin. The nerves of the dorsal row arise from the dorsal branches of the thoracic nerves.

The **lateral thoracic artery, vein** and **nerve** emerge from the axilla between the latissimus dorsi and deep pectoral muscles. The nerve is motor to the cutaneous trunci and may be found on its ventral border. The vessels supply the muscle, skin and subcutaneous tissues including the mammae of the region.

Transect the pectoral muscles close to the sternum. Reflect them toward the forelimb to expose the axilla.

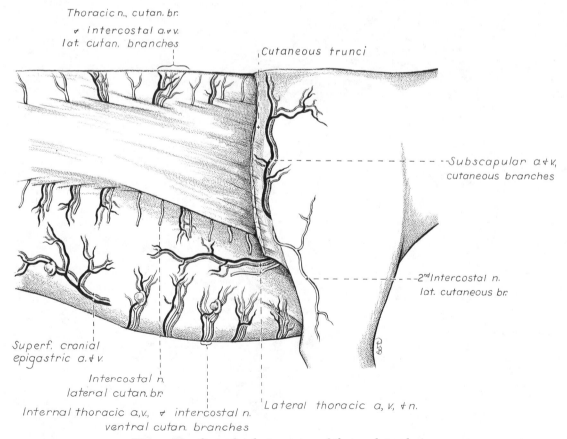

Figure 84. *Superficial structures of thorax, lateral view.*

The **axillary lymph node** lies dorsal to the deep pectoral muscle and caudal to the large vein coming from the arm. Most of the afferent lymph vessels of the thoracic wall and limb drain into this node.

DEEP VESSELS AND NERVES OF THE THORACIC WALL

Expose the costal origin of the external abdominal oblique and detach it. Reflect the muscle ventrally to the rectus abdominis. Free the thoracic attachment of the rectus abdominis close to the sternum and first costal cartilage. Reflect the rectus abdominis caudally, noting any nerves or vessels that enter the deep face of the muscle from any of the intercostal spaces.

On the deep surface of the rectus abdominis the cranial epigastric artery gives rise to the **superficial cranial epigastric,** which perforates the muscle and runs caudally on its surface. This artery supplies the skin over the rectus abdominis and, when present, the cranial abdominal mamma. The cranial epigastric vessels continue on the deep surface of the rectus abdominis. Most of their branches terminate in this muscle.

Make a paramedian incision through the thoracic wall 1 cm. from the midventral line on each side. These incisions should extend from the thoracic inlet through the eighth costal cartilage. The **transversus thoracis** muscle lies on the inner surface of the sternum and costal cartilages. Connect the caudal ends of right and left paramedian incisions and free the sternum, except for the wide thin fold of mediastinum which is now its only attachment.

On the right half of the thorax clean and transect the origin of the latissimus dorsi and reflect it toward the forelimb. Locate and transect the caudal portion of the origin of the serratus ventralis, exposing the ribs. Starting at the costal arch, using bone forceps, cut the ribs close to their origin. Reflect the thoracic wall. As this is done cut the attachments of the internal abdominal oblique, transversus abdominis and diaphragm from the ribs along the costal arch. Reflect the left thoracic wall in a similar manner.

On the internal surface of the thoracic wall notice the intercostal vessels and nerves coursing along the caudal border of the ribs. Ventrally the vessels bifurcate and anastomose with the intercostal branches of the internal thoracic artery and vein. The intercostal nerves supply the intercostal musculature. Their sensory branches were seen as lateral and ventral cutaneous branches.

The **pleurae** (Figs. 85, 86) are serous membranes which cover the lungs and line the walls of the thorax. These form right and left sacs which enclose the pleural cavities.

The **pulmonary pleura** closely attaches to the surfaces of the lungs, following all their small irregularities as well as the fissures which divide the organ into lobes.

The **parietal pleura** is attached to the thoracic wall by the endothoracic fascia. This pleura may be divided into costal, diaphragmatic and mediastinal parts. Each of these is named after the region or surface it covers and all are continuous, one with another. The **costal pleura** covers

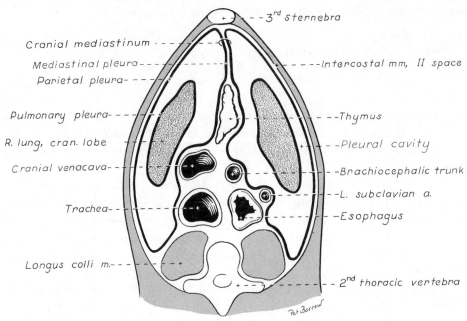

Figure 85. *Schematic cross-section of thorax through cranial mediastinum.*

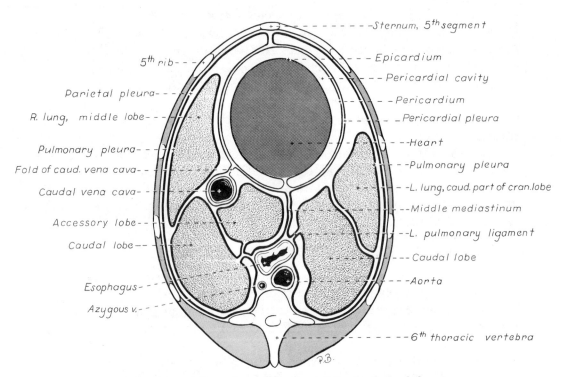

Figure 86. *Schematic cross-section of thorax through heart.*

the inner surfaces of the ribs and intercostal muscles. The **mediastinal pleurae** are the layers which cover the sides of the partition between the two pleural cavities. The **mediastinum** includes the two mediastinal pleurae and the space between them. Enclosed in the mediastinum are the thymus, the lymph nodes, the heart, the aorta, the trachea, the esophagus, the vagus nerves and other nerves and vessels. The **pericardial mediastinal pleura** is that portion covering the heart.

The **mediastinum** can be divided into a cranial part, that lying cranial to the heart; a middle part, that containing the heart; and a caudal part, that lying caudal to the heart. The caudal mediastinum is thin. It attaches to the diaphragm far to the left of the median plane. Cranially it is continuous with the middle mediastinum.

The **plica vena cava** is a loose fold of pleura derived from the right pleural sac which surrounds the caudal vena cava. The **root** of the lung is composed of pleura along with the bronchi, vessels and nerves entering the lung. Here the mediastinal parietal pleura is continuous with the pulmonary pleura. Caudal to the hilus this connection forms a free border between the caudal lobe of the lung and the mediastinum at the level of the esophagus known as the **pulmonary ligament.** Observe this ligament.

The **thymus** (Figs. 87, 88) is a bilobed, compressed structure situated in the cranial mediastinum. It is largest in the young dog and usually atrophies with age until only a trace remains. When maximally developed the caudal part of the thymus is molded on the cranial surface of the pericardium.

The **internal thoracic artery** leaves the subclavian artery, courses ventrocaudally in the cranial mediastinum and disappears deep to the cranial border of the transversus thoracis muscle. It supplies many branches to surrounding structures—the phrenic nerve, the thymus, the

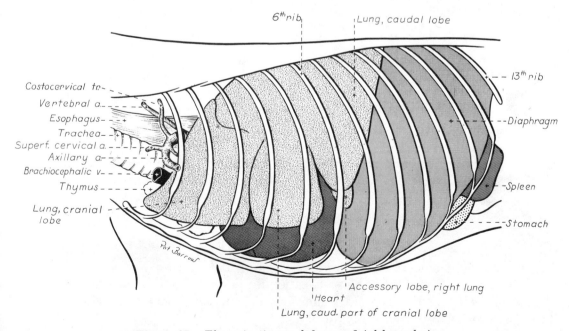

Figure 87. Thoracic viscera, left superficial lateral view.

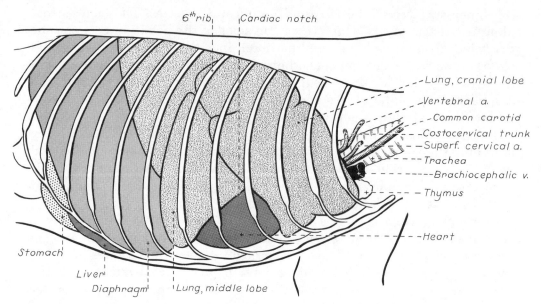

Figure 88. *Thoracic viscera, right superficial lateral view.*

mediastinal pleurae and the intercostal spaces. The ventral cutaneous branches to the superficial structures of the ventral third of the thorax have been seen. The anastomoses with the intercostal arteries on the medial side of the thoracic wall have been seen. The internal thoracic artery terminates in the musculophrenic artery and the larger cranial epigastric artery. The latter has been dissected along with its cranial superficial epigastric branch. The **musculophrenic artery** runs caudodorsolaterally in the angle formed by the diaphragm and lateral thoracic wall. Dissect its origin. Cut the mediastinum and reflect the sternum cranially.

THE LUNGS

The left lung (Figs. 87–89) is divided into **cranial** and **caudal lobes** by deep fissures. The cranial lobe is further divided into cranial and caudal parts. The right lung is divided into **cranial, middle, caudal,** and **accessory lobes.** A part of the accessory lobe can be seen through either the caudal mediastinum or the plica venae cavae.

Examine the **cardiac notch** of the right lung at the fourth and fifth intercostal spaces. The apex of the notch is continuous with the fissure between cranial and middle lobes. A larger area of the ventral convexity of the heart is exposed on the right side. The right ventricle occupies this area of the heart.

Remove the lungs by transecting all structures that enter the hilus. On the right side this will involve slipping the accessory lobe over the caudal vena cava. Make the transection far enough from the heart so that nerves crossing the heart are not severed but close enough so that the lobes are not removed individually.

Examine the structures which attach the lungs. Notice that there is usually a single pulmonary vein from each lobe which drains directly into the left atrium of the heart. (The pulmonary veins contain red latex because the specimen was prepared by injecting the latex into the carotid artery. In a retrograde direction this in turn filled the aorta, left ventricle, left atrium and pulmonary veins. Because latex does not cross capillary beds there is usually no latex in the pulmonary arteries. Occasionally the pressure of injection ruptures the interatrial or interventricular septum in the heart, flooding the right chambers with the latex and thus filling the pulmonary arteries as well as the veins.)

The pulmonary trunk supplies each lung with a pulmonary artery. The left pulmonary artery lies cranioventral to the left bronchus. The artery and bronchus are at a more dorsal level than the veins. Using a scissors or scalpel, open a few of the major bronchi to observe the lumen.

Note the **tracheobronchial lymph nodes** located at the bifurcation of the trachea and also farther out on the bronchi.

Determine by replacing the lungs in the thorax which structures form the various grooves and impressions. Observe the long **aortic impression** of the left lung. The most marked impressions on the right lung are on the accessory lobe. This lobe is interposed between the caudal vena cava on one side and the esophagus on the other, and both leave impressions on it. Observe the vascular impressions on the cranial lobes of the lungs and the costal impressions on each lung.

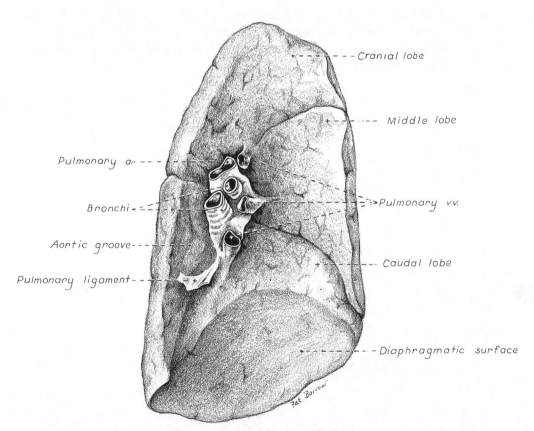

Figure 89. Left lung, medial view.

VEINS CRANIAL TO THE HEART

The **cranial vena cava** (Figs. 90-92) drains into the right atrium after its formation by the union of the right and left brachiocephalic veins at the thoracic inlet. The **brachiocephalic vein** is formed on each side by the **external jugular** and **subclavian veins.** Sometimes the last branch entering the cranial vena cava is the **azygos vein** (Fig. 90). This usually enters the right atrium directly. It is seen from the right in the mediastinal space winding ventrocranially around the hilus of the right lung. It originates dorsally in the abdomen and collects all of the dorsal intercostal veins on each side as far cranially as the third or fourth intercostal space.

The **thoracic duct** is the chief channel for the return of lymph from lymphatic capillaries and ducts to the venous system. It begins in the sublumbar region between the crura of the diaphragm as a cranial continuation of the **cisterna chyli.** This latter is a dilated structure which receives the lymph drainage from the viscera and pelvic limbs. The thoracic duct runs cranially on the right dorsal border of the thoracic aorta and the ventral border of the azygos vein to the level of the sixth vertebra. Here it crosses the ventral surface of the fifth thoracic vertebra and can be seen on the left side of the middle mediastinal pleura. It continues cranioventrally through the cranial mediastinum to the left brachiocephalic vein, where it usually terminates. The thoracic duct also receives the lymph drainage from the left thoracic limb and the **left tracheal duct** (from the left side of the head and neck). The lymph drainage from the right thoracic limb and the **right tracheal duct** (from the right side of the head and neck) enters the venous system in the vicinity of the right brachiocephalic vein. There are often multiple terminations of a complicated nature which may include swellings or anastomoses.

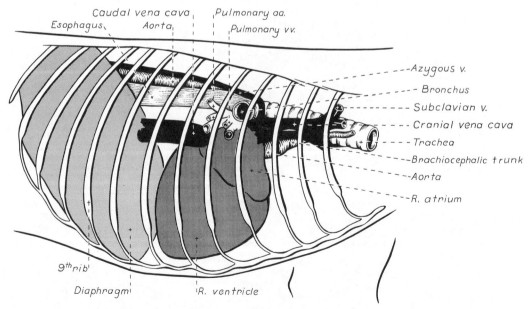

Figure 90. *Thoracic viscera, right deep lateral view.*

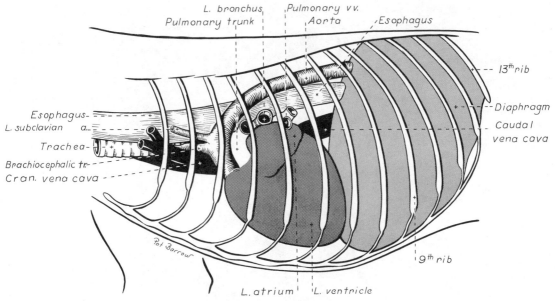

Figure 91. *Thoracic viscera, left deep lateral view.*

Observe the thoracic duct. It may be identified by the reddish-brown or straw color of its contents and the numerous constrictions in its wall. The tracheal ducts may be found in each carotid sheath.

ARTERIES CRANIAL TO THE HEART

The **aorta** (Figs. 92-94) is the large, unpaired vessel which emerges from the left ventricle medial to the pulmonary trunk. As the **ascending aorta** it extends cranially covered by the pericardium; it makes a sharp bend dorsally and to the left as the **aortic arch;** it runs caudally as the **descending aorta** located ventral to the vertebrae. The part cranial to the diaphragm is the thoracic aorta and the caudal part is the abdominal aorta. Cranial to the heart are several branches of the aorta:

The right and left **coronary arteries** are branches of the ascending aorta that supply the heart muscle. They will be studied with the heart.

The **brachiocephalic trunk** (Figs. 92-94), the first branch from the aortic arch, passes obliquely to the right across the ventral surface of the trachea. It gives rise to the **left common carotid artery** and terminates as the **right common carotid artery** and the **right subclavian artery.**

The **left subclavian artery** (Figs. 92, 93) originates from the aortic arch beyond the level of the brachiocephalic.

The branches of the right and left subclavian arteries are similar; only the right subclavian artery will be described. It has four branches that arise medial to the first rib or intercostal space. Do not sever the nerves or arteries.

The **vertebral artery** (Figs. 93, 94) crosses the medial surface of the first rib and disappears between the longus colli and the scalenus muscles. It passes through the transverse foramina of the first six

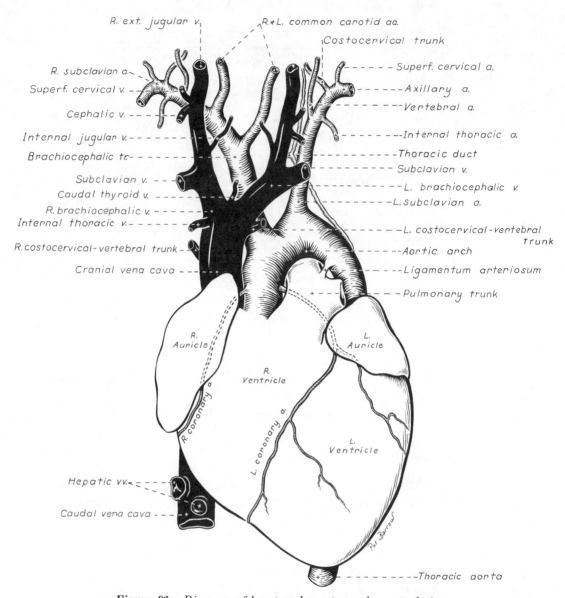

Figure 92. *Diagram of heart and great vessels, ventral view.*

cervical vertebrae. It supplies muscular branches to the cervical muscles, and spinal branches at each intervertebral foramen to the spinal cord and its coverings. At the level of the atlas it terminates by entering the vertebral canal and contributing to the ventral spinal and basilar arteries. These will be seen later in the dissection of the nervous system.

The **costocervical trunk** (Figs. 93, 94) arises distal to the vertebral artery, crosses its lateral side and extends dorsally as far as the vertebral end of the first rib. By its various branches it supplies the structures of the first, second, and third intercostal spaces, the muscles at the base of the neck and the muscles dorsal to the first few thoracic vertebrae. These need not be dissected.

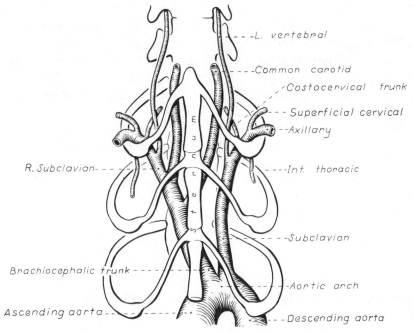

Figure 93. *Branches of aortic arch, ventral view.*

The **superficial cervical artery** (Figs. 93, 94) arises from the subclavian opposite the origin of the internal thoracic artery medial to the first rib. Its branches supply the base of the neck and the adjacent scapular region.

The **internal thoracic artery** has been studied.

BRANCHES OF THE THORACIC AORTA

The **esophageal** and **bronchial arteries** vary in number and origin. Usually the small **bronchoesophageal artery** leaves the right fifth inter-

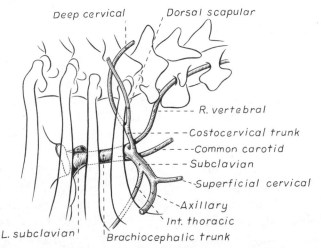

Figure 94. *Branches of brachiocephalic trunk, right lateral view.*

costal artery close to its origin and crosses the left face of the esophagus which it supplies. It terminates shortly afterwards in the **bronchial arteries** which supply the lung.

There are nine pairs of dorsal **intercostal arteries** which leave the aorta (Fig. 81). These start with either the fourth or the fifth intercostal artery and continue caudally, there being an artery in each of the remaining intercostal spaces.

The **phrenic nerves** (Fig. 95) supply the diaphragm. Find each nerve as it passes through the thoracic inlet. The nerve arises from the ventral branches of the fifth, the sixth and usually the seventh cervical nerves. Follow the phrenic nerves to the diaphragm. Each is both motor and sensory to the corresponding half of the diaphragm except at its periphery. This part of the muscle receives sensory fibers from the caudal few intercostal nerves.

INTRODUCTION TO THE AUTONOMIC NERVOUS SYSTEM

The **nervous system** is highly organized on both an anatomical and a functional level. It is composed of a **central nervous system** and a **peripheral nervous system.** The central nervous system includes the **brain** and the **spinal cord.** The peripheral nervous sytem comprises the **cranial nerves** which connect the base of the brain (brainstem) with structures of the head and body, and the **spinal nerves** which connect the spinal cord to structures of the body and extremities.

The peripheral nervous system can be further subdivided on an anatomical and functional basis. The peripheral nerves contain axons that conduct impulses *to* the central nervous system — **sensory, afferent axons;** and axons that conduct impulses *from* the central nervous system to muscles and glands of the body — **motor, efferent axons.** Most peripheral nerves have both sensory and motor axons.

The **motor portion** of the peripheral nervous system is classified according to the type of tissue being innervated. Motor neurons supplying voluntary, striated, skeletal muscle are **general somatic efferent** neurons. Those supplying involuntary, smooth muscle of viscera, and blood vessels, cardiac muscle and glands, are **general visceral efferent** neurons. One specialized group of motor neurons supplies voluntary striated musculature of the head region that is derived from branchial arch mesoderm. These are called **special visceral efferent** neurons.

A **neuron** is composed of a cell body and its processes. Ordinarily a motor neuron of the peripheral nervous system has its cell body located in the gray matter of the spinal cord (or brainstem), and its process or axon courses through the peripheral spinal (or cranial) nerve to end in the muscle innervated. Thus there is only one neuron spanning the distance from the central nervous sytem to the innervated structure.

The general visceral efferent system is the peripheral motor part of the **autonomic nervous system.** It differs anatomically from the other peripheral motor systems in having a second neuron interposed between the central nervous system and the innervated structures. One neuron has its cell body located in the gray matter of the central nervous system. Its axon courses in the peripheral nerves only part way toward the structure to be innervated. Along the course of the peripheral nerve is a gross enlargement called a **ganglion.** This is a collection of neuronal cell bodies located outside the central nervous system. Autonomic ganglia contain the cell bodies of the second motor neurons in the pathway of the autonomic nervous system. Their axons complete the pathway to the structure being innervated. Because of its relationship to the cell bodies in the autonomic ganglia, the first neuron with its cell body in the central nervous system is called the **preganglionic neuron.** The cell body of the second neuron is in an autonomic ganglion. Its axon is postganglionic. A synapse occurs between these two neurons where the preganglionic axon meets the cell body of the postganglionic axon.

The general visceral efferent system is divided into two subdivisions on the basis of anatomical, pharmacological and functional characteristics. They are the sympathetic and parasympathetic divisions.

In the **sympathetic division** the preganglionic cell bodies are limited to the segments of the spinal cord from approximately the first thoracic to the fifth lumbar segments — the thoracolumbar portion. The cell bodies of postganglionic axons are located in ganglia that are usually only a short distance from the spinal cord. At most of the postganglionic nerve endings of this portion of the autonomic nervous system a humoral transmitter substance, norepinephrine, is released, which causes a response in the structures innervated. The overall effect of this system is to help the body withstand unfavorable environmental conditions or conditions of stress.

In the **parasympathetic division** the preganglionic cell bodies are located in specific nuclei in the brainstem associated with cranial nerves 3, 7, 9, 10 and 11 and in the three sacral segments of the spinal cord — the craniosacral portion. The cell bodies of the postganglionic axons are often located in terminal ganglia on or in the wall of the structure being innervated. Others are found in specifically named ganglia near the innervated structure. At the postganglionic nerve endings a humoral transmitter substance, acetylcholine, is released, which causes a response in the structures innervated. This system is concerned with the normal homeostatic activity of the visceral body functions, the conservation and restoration of body resources and reserves.

The anatomy of the sympathetic division of the peripheral autonomic nervous system requires further description before it is dissected. The preganglionic cell bodies are located in the gray matter of the thoracic and first five lumbar spinal cord segments. Their axons leave the spinal cord along with those of other motor neurons in the ventral rootlets of each of these spinal cord segments. Each **ventral root** unites with the corresponding sensory **dorsal root** at the level of the intervertebral foramen to form the **spinal nerve.** The dorsal branch of the spinal nerve branches off immediately. Just beyond this point a nerve leaves the ventral branch of the spinal nerve, the **ramus communicans.** It courses ventrally to join the **sympathetic trunk** which runs in a craniocaudal direction just lateral to the vertebral column. A ganglion is usually located in the trunk at the point where each ramus joins it. This is the **sympathetic trunk ganglion** and contains the cell bodies of postganglionic axons.

Leaving the caudal thoracic and lumbar portions of the sympathetic trunk are nerves that course into the abdominal cavity, the **splanchnic nerves.** These form plexuses around the main blood vessels of the abdominal organs. Additional sympathetic ganglia are located in association with these plexuses and blood vessels. The cell bodies of sympathetic postganglionic axons are located here. These axons follow the terminal branching of the blood vessels of the abdominal organs to reach the organ innervated.

Each preganglionic sympathetic axon must pass through the ramus communicans of its spinal nerve to reach the sympathetic trunk. Its fate from here is variable and mostly dependent on the structure to be innervated. A few examples will be given.

Smooth muscle of blood vessels, piloerector muscles and sweat glands are innervated by postganglionic sympathetic axons in spinal nerves. The preganglionic axon enters the sympathetic trunk via the ramus communicans. It may synapse in the ganglion where it entered or it may pass up or down the sympathetic trunk a few segments and synapse in the ganglion of that segment. The postganglionic axons then return to the segmental spinal nerve via the ramus communicans, usually of the segment where the synapse occurred. The postganglionic axon then courses with the distribution of the spinal nerve to the smooth muscle and sweat glands. Thus the rami communicantes of spinal cord segments T_1 to L_5 contain both preganglionic and postganglionic axons.

For smooth muscle and glands of the head, the preganglionic axons enter the sympathetic trunk in the cranial thoracic region. Some may synapse in ganglia as they enter the sympathetic trunk. Many others continue as preganglionic axons up the sympathetic trunk in the neck where it courses in the same fascial sheath as the common carotid artery and vagus nerve. At the cranial end of this trunk just below the base of the skull a ganglion is located — the **cranial cervical ganglion.** All remaining preganglionic axons to the head will synapse here. The postganglionic axons are then distributed with the blood vessels to the structures of the head innervated by this sympathetic system.

The sympathetic trunk is located along the full length of the vertebral column on both sides. Throughout the thoracic, lumbar and sacral levels it is joined to each segmental spinal nerve by a ramus communicans. Only those from spinal cord segments T_1 to L_5 contain preganglionic axons.

For smooth muscle and glands in the abdominal and pelvic cavities, the preganglionic sympathetic axon reaches the sympathetic trunk via the ramus communicans. It may synapse with a postganglionic neuron in a trunk ganglion but more often it continues through the ganglion without synapsing and enters a splanchnic nerve. The preganglionic (or occasionally postganglionic) axon courses through the appropriate splanchnic nerve to the abdominal autonomic plexuses and their ganglia. Preganglionic axons synapse in one of these ganglia with a cell body of a postganglionic axon. The postganglionic axons follow the terminal branches of abdominal blood vessels to the organs innervated.

Dissection. Selected portions of the autonomic nervous system will be dissected as they are exposed in the regions being studied.

Examine the dorsal aspect of the interior of the pleural cavities (Fig. 95). Notice the sympathetic trunks coursing longitudinally across the ventral surface of the necks of the ribs. The small enlargements in these trunks at each intercostal space are sympathetic trunk ganglia. Dissect a portion of the trunk and a few ganglia on either side. Notice the fine filaments that run dorsally between the vertebrae to join the spinal nerve of that space. These are the rami communicantes of the sympathetic trunk.

Follow the thoracic portion of the sympathetic trunk cranially. Notice the irregular enlargement of the trunk medial to the dorsal end of the first intercostal space on the lateral side of the longus colli. This is the **cervicothoracic ganglion.** It is formed by a collection of cell bodies from a fusion of caudal cervical and thoracic ganglia. Locate this ganglion on both sides.

Many branches leave the cervicothoracic ganglion. Rami communicantes connect to the first and second thoracic spinal nerves and to the seventh and eighth cervical spinal nerves. These spinal nerves contribute to the formation of the **brachial plexus** and this provides a pathway for postganglionic axons to reach the thoracic limb. A branch or plexus follows the vertebral artery through the transverse foramina — the **vertebral nerve.** This is a source of postganglionic axons to the remaining cervical spinal nerves via branches at each intervertebral space from the vertebral nerve to the cervical spinal nerve. Postganglionic axons may leave the cervicothoracic ganglion and course directly to the heart.

The sympathetic trunk, cranial to the cervicothoracic ganglion, divides and passes dorsal and ventral to the subclavian artery. The ventral part is the **ansa subclavia.** The sympathetic trunk and the ansa reunite and join the **middle cervical ganglion.** This ganglion lies at the junction of the ansa and the vagosympathetic trunk and appears as a swelling of the combined structures. Locate these structures on both sides. Numerous branches, **cardiac nerves,** leave the ansa and middle cervical ganglion and course to the heart.

The **vagosympathetic trunk** in the neck lies in the carotid sheath. Its sympathetic portion carries preganglionic and postganglionic sympathetic axons cranially to structures in the head. The **cranial cervical ganglion** is located at its most cranial end. This is at the level of the base of the ear, just caudomedial to the tympanic bulla. It will be dissected later. The tenth cranial or **vagus nerve** contains parasympathetic preganglionic axons that course caudally down the neck to thoracic and abdominal organs.

At the level of the middle cervical ganglion notice the vagus nerve where it leaves the vagosympathetic trunk to continue its course caudally. Cardiac nerves leave the vagus to innervate the heart. Study the caudal course of the vagus nerve on each side.

At the middle cervical ganglion or slightly caudal to it the **right**

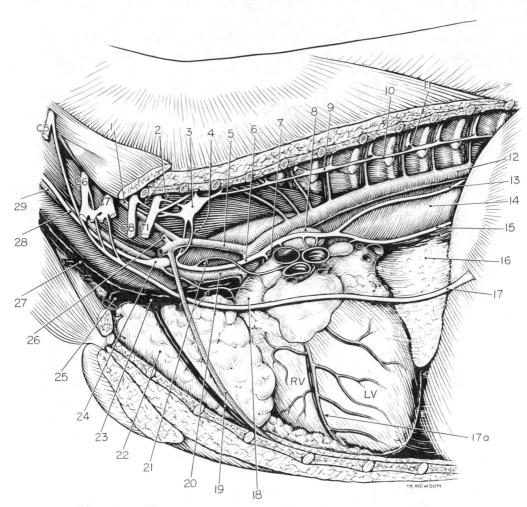

Figure 95. *Thoracic autonomic nerves, lateral view, left lung removed.*

1. Vertebral artery and nerve
2. Communicating rami from cervicothoracic ganglion to ventral branches of cervical and thoracic nerves
3. Left cervicothoracic ganglion
4. Ansa subclavia
5. Left subclavian artery
6. Left vagus nerve
7. Left recurrent laryngeal nerve
8. Left tracheobronchial lymph node
9. Sympathetic trunk ganglion
10. Sympathetic trunk
11. Ramus communicans
12. Aorta
13. Dorsal branch of vagus nerve
14. Esophagus
15. Ventral trunk of vagus nerve
16. Accessory lobe of lung

17. Phrenic nerve to diaphragm
17a. Paraconal inter. groove
18. Pulmonary trunk
19. Internal thoracic artery and vein
20. Brachiocephalic trunk
21. Cardiac autonomic nerves
22. Thymus
23. Cranial vena cava
24. Middle cervical ganglion
25. Left subclavian vein
26. Costocervical trunk
27. External jugular vein
28. Vagosympathetic trunk
29. Common carotid artery

RV – Right ventricle
LV – Left ventricle

recurrent laryngeal nerve leaves the vagus, curves dorsocranially around the right subclavian artery, reaches the dorsolateral surface of the trachea and courses cranially to the larynx. It may be found in the angle between the trachea and the longus colli. On the left side, the left recurrent laryngeal nerve leaves the vagus caudal to the middle cervical ganglion, curves medially around the arch of the aorta and becomes related to the ventrolateral aspect of the trachea and the ventromedial edge of the esophagus. In this position it courses the neck to reach the larynx. As it ascends it reaches the dorsolateral aspect of the trachea. Each recurrent nerve sends branches to the heart, trachea and esophagus before terminating in the laryngeal muscles. The laryngeal nerves will be dissected later.

Follow each vagus nerve as it courses over the base of the heart and supplies cardiac nerves to it. Branches are supplied to the bronchi as the vagus passes over the roots of the lungs. Between the azygos vein and the right bronchus on the right and the area just caudal to the base of the heart on the left, each vagus divides into dorsal and ventral branches. The right and left ventral branches soon unite with each other to form the ventral vagal trunk on the esophagus. The dorsal branches of each vagus do not unite until farther caudally where they form the dorsal vagal trunk, which lies dorsal to the esophagus. The termination of these trunks in the abdomen will be studied later.

HEART AND PERICARDIUM

The pericardium is the fibroserous covering of the heart, which is located in the middle part of the mediastinum from the level of the third to the sixth rib. The pericardial mediastinal pleura covers the pericardium. The epicardium is a serous membrane which covers the heart muscle. The pericardium and epicardium are continuous at the base of the heart. Between these layers is the pericardial cavity, which contains a small amount of pericardial fluid. Incise the combined pericardial mediastinal pleura and pericardium from the base to the apex of the heart and reflect it.

The heart (Figs. 96, 97) has two surfaces. The caudal surface which faces the diaphragm is the diaphragmatic surface. The sternocostal surface faces cranioventrally and is apposed to the sternum and ribs. Notice that the thin-walled right ventricle winds across the whole sternocostal surface but that the greater part of it lies to the right of the median plane.

Trace the coronary groove around the heart. It lies between the atria and ventricles and contains coronary vessels and fat. The interventricular grooves are the superficial separations of the right and left ventricles. The subsinual interventricular groove is a short furrow which lies on the diaphragmatic surface and marks the approximate position of the interventricular septum. Obliquely traversing the sternocostal surface of the heart is the paraconal interventricular groove. Longer and more distinct than the subsinual, it begins at the base of the pulmonary trunk where it is covered by the left auricle.

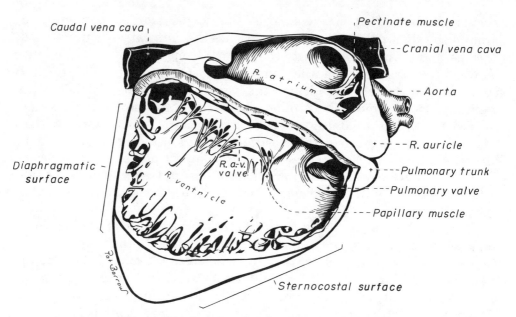

Figure 96. *Interior of right atrium and ventricle.*

The **right atrium** receives the blood from the systemic veins and most of the blood from the heart itself. It lies dorsocranial to the right ventricle. It is divided into a main part, the **sinus venarum,** and a blind cranial part, the **right auricle.**

Open the right atrium by a longitudinal incision through its wall from the cut end of the cranial vena cava to the caudal vena cava. Extend a cut from the middle of the first incision to the tip of the auricle.

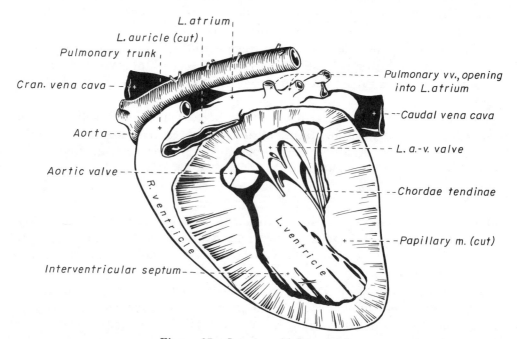

Figure 97. *Interior of left ventricle.*

There are four openings into the sinus venarum of the right atrium. The **caudal vena cava** enters the atrium from behind. Ventral to this opening is the **coronary sinus,** the enlarged venous return for most of the blood from the heart. The **cranial vena cava** enters the atrium from above and cranially. Ventral and cranial to the coronary sinus is the large opening from the right atrium to the right ventricle, the **right atrioventricular orifice.** The valve will be described with the right ventricle.

Examine the dorsomedial wall of the sinus venarum, the **interatrial septum.** Between the two caval openings is a transverse ridge of tissue, the **intervenous tubercle.** It diverts the inflowing blood from the two caval veins toward the right atrioventricular orifice. Caudal to the intervenous tubercle is a slit-like depression, the **fossa ovalis.** In the fetus, there is an opening at the site of the fossa, the foramen ovale, which allows blood to pass from the right to the left atrium.

The **right auricle** is the blind, ear-shaped pouch of the right atrium which faces cranially. The internal surface of the wall of the right auricle is strengthened by interlacing muscular bands, the **pectinate muscles.** These are also found on the lateral wall of the atrium proper. The internal surface of the heart is lined everywhere with a thin, glistening membrane, the **endocardium,** which is continued in the blood vessels as the tunica intima.

Locate the **pulmonary trunk** leaving the right ventricle at the left craniodorsal angle of the heart. Begin at the cut end of the left pulmonary artery and extend an incision through the wall of this artery, the pulmonary trunk and the wall of the right ventricle along the paraconal interventricular groove. Continue this cut around the right ventricle following the interventricular septum to the origin of the subsinual interventricular groove. Reflect the ventricular wall.

The pulmonary trunk bifurcates into right and left pulmonary arteries, each going to its respective lung. At the junction between the pulmonary trunk and the right ventricle is the **pulmonary valve,** consisting of three semilunar cusps. A small, fibrous **nodule** is located at the middle of the free concave margin of each semilunar cusp.

The greater part of the base of the right ventricle communicates with the right atrium through the atrioventricular orifice. This opening is provided with the **right atrioventricular valve.** There are two main parts to the valve in the dog: a wide but short flap which arises from the parietal margin of the orifice, the **parietal cusp,** and a flap from the septal margin, the **septal cusp,** which is nearly as wide as it is long. Subsidiary leaflets are found at each end of the septal flap. The points of the flaps of the valve are continued to the septal wall of the ventricle by the **chordae tendineae.** The chordae tendineae are attached to the septal wall by means of conical muscular projections, the **papillary muscles,** of which there are usually three to four. The **trabeculae carneae** are the muscular irregularities of the interior of the ventricular walls. The **trabecula septomarginalis** is a muscular strand which extends across the lumen of the ventricle from the septal to the marginal wall. The right ventricle terminates as the funnel-shaped **conus arteriosus** which gives rise to the pulmonary trunk.

The **left atrium** is situated on the left dorsocaudal part of the base of the heart dorsal to the left ventricle. Open the left atrium by a longitudinal incision through its wall from the apex of the auricle to the lumen of the farthest left pulmonary vein. Five or six openings mark the entrance of the pulmonary veins into the atrium. The inner surface of the atrium is smooth except for pectinate muscles confined to the **left auricle.** A thin flap of tissue is present on the cranial part of the interatrial septal wall. This is the **valve of the foramen ovale.**

Remove the fat, pleura and pericardium from the aorta. In doing this isolate the **ligamentum arteriosum,** a fibrous connection between the pulmonary trunk and the aorta. In the fetus it was patent (**ductus arteriosus**) and served to shunt the blood destined for the non-functional lungs to the aorta. Isolate the origins of the pulmonary trunk and aorta.

To open the left ventricle and aorta make a longitudinal incision through the wall of the left ventricle. Begin at the apex on the left lateral side of the heart about midway between the subsinual and paraconal interventricular grooves. Continue this incision dorsally through the left atrioventricular valve and left atrium to the incision which opened the left atrium.

Notice the thickness of the left ventricular wall as compared with the right. The **left atrioventricular valve** is composed of two major cusps, **septal** and **parietal,** but the division is indistinct. Secondary cusps are present at the ends of the two major ones. Notice the papillary muscles and their chordae tendinae attached to the cusps. The trabeculae carneae are not as numerous in the left ventricle as in the right.

Insert scissors into the aortic valve located beneath the septal cusp of the left atrioventricular valve and cut the aortic valve, aortic wall and left atrium to expose the aortic valve and the first centimeter of the ascending aorta. The **aortic valve,** like the pulmonary, consists of three semilunar cusps. Notice the nodules of the semilunar cusps in the middle of their free borders. Behind each cusp, the aorta is slightly expanded to form the **sinus** of the aorta.

The **right coronary artery** (Fig. 92) leaves the right sinus of the aorta. It encircles the right side of the heart in the coronary groove and often extends to the subsinual interventricular groove. It sends many small and one or two large descending branches over the surface of the right ventricle. Remove the epicardium and fat from its surface and follow the artery to its termination.

The **left coronary artery** (Fig. 92) is about twice as large as the right. It is a short trunk that leaves the left sinus and immediately terminates in (1) a **circumflex branch** which extends caudally in the left part of the coronary groove, and (2) a **paraconal interventricular branch** which obliquely crosses the sternocostal surface of the heart in the paraconal interventricular groove. Both of these branches send large rami over the surface of the left ventricle. Expose the artery and its large branches by removing the epicardium and fat.

The **coronary sinus** is the dilated terminal end of the great cardiac vein. The **great cardiac vein,** which begins in the paraconal interventric-

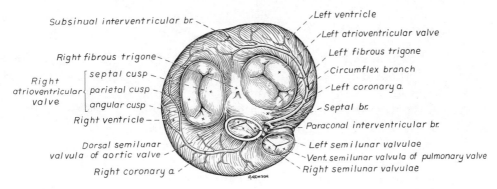

Figure 98. *Atrioventricular, aortic and pulmonary valves, craniodorsal view.*

ular groove, returns blood supplied to the heart by the left coronary artery. Clean the surface of the great cardiac vein and open the coronary sinus. Usually one or two poorly developed valves are present in the coronary sinus.

VESSELS AND NERVES OF THE THORACIC LIMB

The main artery to the thoracic limb arises within the thorax as a terminal branch of the brachiocephalic on the right side, and directly from the aorta on the left side. It is divided into three parts. That which extends from its origin to the first rib is the **subclavian artery** (Fig. 95). From the first rib to the conjoined tendons of the teres major and latissimus dorsi the vessel is the **axillary artery.** From here to the terminal median artery distal to the elbow the vessel is called the **brachial artery.**

The **superficial cervical artery** (Figs. 92, 94) arises from the subclavian just inside the thoracic inlet. It runs dorsocranially between the scapula and the neck. It supplies the superficial muscles of the base of the neck, the superficial cervical lymph nodes, the muscles of the scapula and the shoulder.

There are usually two **superficial cervical lymph nodes** which lie cranial to the supraspinatus covered by the omotransversarius and the brachiocephalicus. These nodes receive the afferent lymph vessels from the superficial part of the lateral surface of the neck, the convex surface of the ear and the lateral surface of the thoracic limb.

BRACHIAL PLEXUS

The brachial plexus (Figs. 100, 101) is formed by the ventral branches of the sixth, seventh and eighth cervical and the first and second thoracic spinal nerves. These branches pass between vertebrae, emerge along the ventral border of the scalenus and extend across the axillary space to the thoracic limb. In the axilla numerous branches of these nerves communicate to form the brachial plexus. From the plexus arise nerves of mixed origin which supply the structures of the thoracic limb and adjacent muscles and skin (Fig. 109b).

The pattern of interchange in the brachial plexus is variable but the specific spinal nerve composition of the named nerves that continue into the thoracic limb is consistent. These nerves include the suprascapular, subscapular, axillary, musculocutaneous, radial, median, ulnar, thoracodorsal, lateral thoracic and pectoral nerves. Expose the brachial plexus in the axilla.

Reflect the superficial and deep pectoral muscles toward their insertions to expose the vessels and nerves on the medial aspect of the arm.

AXILLARY ARTERY

The axillary artery (Figs. 101, 102) is the continuation of the subclavian which extends from the first rib to the conjoined tendons of the teres

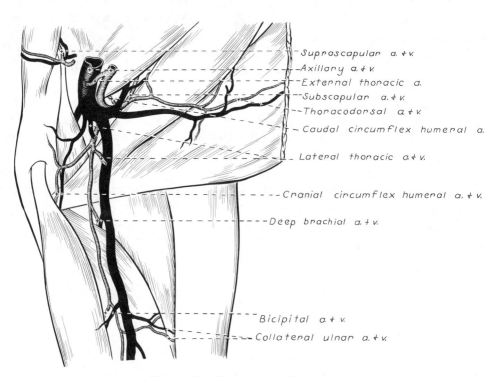

Figure 99. Vessels of axillary region.

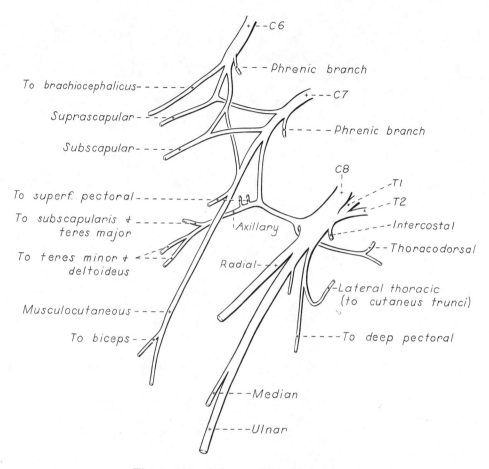

Figure 100. *Schema of brachial plexus.*

major and latissimus dorsi. It has four branches: the external thoracic, the lateral thoracic, the subscapular and the cranial circumflex humeral. In some specimens the origin of the arteries will differ from the description. In such cases identify the vessel by its distribution.

1. The **external thoracic artery** (Figs. 99, 102) leaves the axillary near its origin. The external thoracic artery curves around the craniomedial border of the deep pectoral with the nerve to the superficial pectorals and is distributed almost entirely to the superficial pectorals. It may arise from a common trunk with the lateral thoracic artery or it may arise from the deltoid branch of the superficial cervical artery.

2. The **lateral thoracic artery** (Fig. 103) runs caudally across the lateral surface of the axillary lymph node. It usually arises from the axillary artery distal to the external thoracic. The vessel may arise distal to the subscapular artery. It supplies parts of the latissimus dorsi, deep pectoral and cutaneus trunci muscles and the thoracic mammae.

3. The **subscapular artery** (Figs. 99, 102) is larger than the continu-

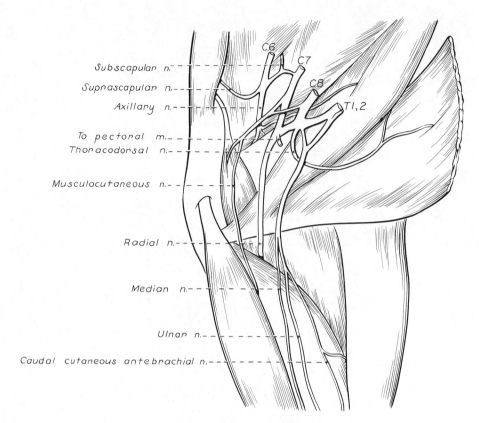

Figure 101. *Nerves of axillary region.*

ation of the axillary in the arm. Only a short part of the subscapular is now visible. It passes between the subscapularis and the teres major and becomes subcutaneous near the caudal angle of the scapula. Dissect the following branches:

The **thoracodorsal artery** (Figs. 99, 102) leaves the dorsal surface of the subscapular near its origin. It supplies a part of the teres major and latissimus dorsi and ends in the skin. Transect the teres major and reflect both ends to expose the subscapular artery.

The **caudal circumflex humeral artery** (Figs. 99, 102) leaves the subscapular opposite the thoracodorsal and courses between the head of the humerus and the teres major. Expose the caudal circumflex humeral artery from the lateral side where branches become superficial cranial to the dorsal part of the lateral head of the triceps. Remove the skin from the lateral surface of the shoulder and arm. Transect the insertion of the deltoideus. Reflect the deltoideus proximally and observe the axillary nerve and caudal circumflex humeral artery entering the deep surface of the muscle. Notice the large branch of the cephalic vein which travels with the artery and nerve. Transect the long head of the triceps at its origin. This artery supplies the triceps, deltoideus, coracobrachialis and infraspinatus muscles and the shoulder.

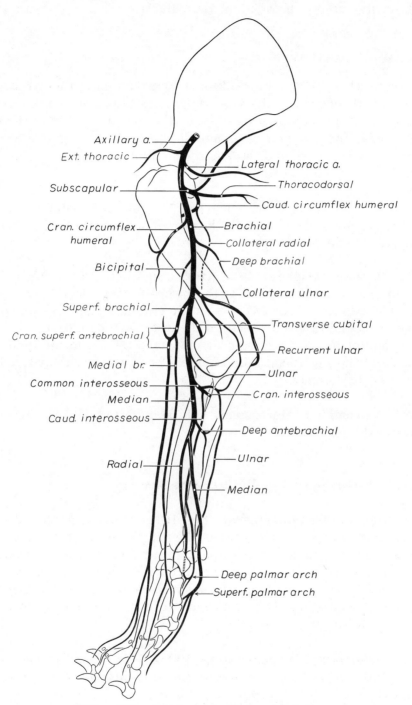

Figure 102. *Arteries of forelimb, medial view.*

Follow the continuation of the subscapular artery caudodorsally along the caudal surface of the scapula. Numerous muscular branches supply the adjacent musculature.

BRACHIAL ARTERY

The brachial artery (Fig. 102) is a continuation of the axillary from the conjoined tendons of the teres major and latissimus dorsi. It crosses the distal half of the humerus to reach the medial surface of the elbow, passes under the pronator teres muscle, gives off its largest branch, the **common interosseous,** and becomes the **median artery.** Dissect the following branches of the brachial artery:

1. The **collateral ulnar artery** (Fig. 102) is a caudal branch of the brachial in the distal third of the arm. It supplies the triceps, the ulnar nerve and the elbow.

2. The **superficial brachial artery** (Fig. 102) loops around the cranial surface of the distal end of the biceps brachii, deep to the cephalic vein. It continues in the forearm as the cranial superficial antebrachial artery. A medial ramus arises from the latter and both course distally on either side of the cephalic vein accompanied by the medial and lateral branches of the superficial radial nerve. These vessels supply blood to the dorsum of the forepaw.

3. The **common interosseous** will be dissected with the vessels and nerves of the forearm.

NERVES OF THE SCAPULAR REGION AND ARM

1. The **suprascapular nerve** (Fig. 101) leaves the sixth and seventh cervical nerves and courses between the supraspinatus and subscapularis muscles near the neck of the scapula. It supplies the supraspinatus and infraspinatus. Transect the supraspinatus at its insertion and reflect the distal end. Trace the branches of the suprascapular nerve to this muscle. Note the continuation of the nerve distal to the scapular spine where it enters the infraspinatus muscle.

2. The **subscapular nerve** (Figs. 100, 101) is a branch from the sixth and seventh cervicals to the subscapularis. Sometimes two nerves enter the muscles.

3. The **musculocutaneous nerve** (Figs. 101, 104) arises from the sixth, seventh, and eighth cervical nerves. Throughout the brachium the musculocutaneous nerve lies between the biceps brachii cranially and the main vessels caudally. It supplies the coracobrachialis, the biceps brachii and the brachialis. A branch communicates with the median nerve proximal

to the flexor surface of the elbow. The musculocutaneous nerve courses deep to the insertion of the biceps. It supplies the distal end of the brachialis and gives off the **medial cutaneous antebrachial nerve.**

4. The **axillary nerve** (Figs. 100, 101) arises as a branch from the combined seventh and eighth cervical nerves. It enters the space between the subscapularis and the teres major on a level with the neck of the

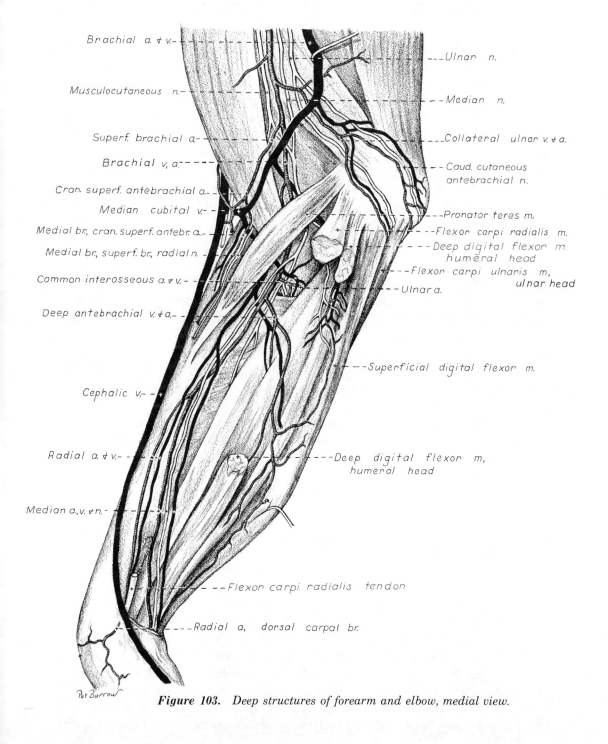

Figure 103. *Deep structures of forearm and elbow, medial view.*

scapula. The following muscles are supplied by the axillary nerve: the teres major, the teres minor, the deltoideus and the subscapularis in part. The cranial lateral cutaneous brachial nerve appears subcutaneously on the lateral head of the triceps, where it supplies the skin of the lateral surface of the brachium and the proximal lateral aspect of the forearm.

5. The **thoracodorsal nerve** (Figs. 100, 101) arises primarily from the eighth cervical nerve. It supplies the latissimus dorsi muscle. It courses with the thoracodorsal vessels on the medial surface of the muscle.

6. The **radial nerve** (Figs. 100, 101) arises from the last two cervical and first two thoracic nerves, runs a short distance distally with the trunk of the median and ulnar nerves and enters the triceps distal to the teres major. The radial nerve is motor to all the extensor muscles of the elbow, carpal and phalangeal joints. The muscles of the arm supplied by the radial nerve are the triceps, the tensor fasciae antebrachii and the anconeus. Observe the branches to the triceps. The radial nerve spirals around the humerus in relation to the caudal and then lateral surface of the brachialis muscle. On the lateral side at the distal third of the arm the radial nerve terminates as a **deep** and a **superficial branch.** Transect the lateral head of the triceps at its origin and reflect it to expose these terminal branches.

7. The **median** and **ulnar nerves** (Figs. 101, 104, 105) arise by a common trunk from the eighth cervical and the first and second thoracic nerves. The common trunk lies on the medial head of the triceps between the brachial vein caudally and the brachial artery cranially. The **median nerve,** the cranial division of the common trunk, runs to the antebrachium in contact with the caudal surface of the brachial artery. It receives a branch from the musculocutaneous nerve at the level of the elbow. Its branches to several of the muscles of the forearm and to the skin of the palmar side of the paw will be dissected later.

The **ulnar nerve,** the caudal division of the common trunk, crosses the elbow caudal to the medial epicondyle of the humerus. Trace it with the collateral ulnar artery to the cut edge of the skin. The **caudal cutaneous antebrachial nerve** (Fig. 105) leaves the ulnar near the middle of the arm and runs caudodistally across the medial surface of the triceps and olecranon.

Incise the skin from the olecranon to the palmar surface of the third digit. (Pass through the carpal, metacarpal and third digital pads.) Remove the skin from the forearm and paw, leaving the vessels and nerves on the specimen.

The **cephalic vein** (Figs. 80, 103) begins on the palmar side of the paw from the superficial palmar venous arch. This need not be dissected. The **accessory cephalic vein** joins the cephalic on the cranial surface of the distal third of the forearm. It arises from small veins on the dorsum of the paw. At the flexor surface of the elbow the **median cubital vein** forms a connection between the brachial and the cephalic veins. From this connection the cephalic continues proximally on the craniolateral surface of the arm in a furrow between the brachiocephalicus cranially and the origin of

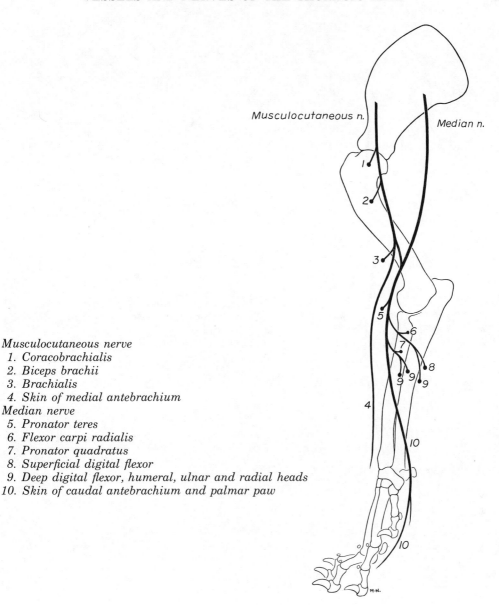

Musculocutaneous n.

Median n.

Musculocutaneous nerve
1. Coracobrachialis
2. Biceps brachii
3. Brachialis
4. Skin of medial antebrachium
Median nerve
5. Pronator teres
6. Flexor carpi radialis
7. Pronator quadratus
8. Superficial digital flexor
9. Deep digital flexor, humeral, ulnar and radial heads
10. Skin of caudal antebrachium and palmar paw

Figure 104. Distribution of musculocutaneous and median nerves, medial view.

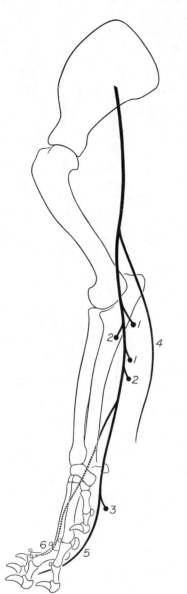

1. *Flexor carpi ulnaris, ulnar and humeral heads*
2. *Deep digital flexor, ulnar and humeral heads*
3. *Interossei*
4. *Skin of caudal antebrachium*
5. *Skin of palmar paw*
6. *Skin of fifth metacarpal digit, lateral surface*

Figure 105. *Distribution of ulnar nerve, medial view.*

the lateral head of the triceps caudally. In the middle of the arm a branch, the **cephalic vein,** runs deep to the brachiocephalicus and enters the external jugular. The axillobrachial vein passes deep to the caudal border of the deltoideus to join the axillary vein. The omobrachial vein arises from the axillobrachial vein and continues subcutaneously across the cranial surface of the arm and shoulder before entering the external jugular.

ARTERIES OF THE FOREARM AND PAW

Make a longitudinal incision through the medial part of the antebrachial fascia midway between the cranial and caudal borders. Extend this incision to the carpus. Remove the fascia from the forearm. Transect the pronator teres close to its origin and reflect it to uncover the brachial artery.

The **brachial artery** (Fig. 102) in the forearm gives rise to the common interosseous artery and continues as the median artery. The median extends from the origin of the common interosseous artery to the superficial palmar arch in the paw.

1. The **common interosseous artery** (Fig. 102) passes to the proximal part of the interosseous space between the radius and ulna.

The **ulnar artery** courses caudally. Separate the muscles on the caudomedial side of the forearm to expose its course. A recurrent branch extends proximally between the humeral and ulnar heads of the deep digital flexor. The ulnar artery continues distally with the ulnar nerve between the humeral head of the deep digital flexor and the flexor carpi ulnaris. It supplies the ulnar and humeral heads of the deep digital flexor and the flexor carpi ulnaris.

The **caudal interosseous artery** lies between the apposed surfaces of the radius and ulna. On the medial side of the forearm expose the pronator quadratus muscle between the radius and ulna. Cut the attachments of this muscle between the two bones and remove it from the interosseous space. The caudal interosseous artery lies deep in this space. In its course down the forearm the artery supplies many small branches to adjacent structures. At the carpus the caudal interosseous artery joins with branches of the radial and median arteries to form arches which supply the palmar surface of the forepaw. These will not be dissected.

The **cranial interosseous artery** passes through the proximal part of the interosseous space cranially to supply the muscles lying laterally and cranially in the forearm.

2. The **median artery** (Fig. 102) is the continuation of the brachial artery beyond the origin of the common interosseous. The deep antebrachial and radial arteries branch off the median in the middle of the forearm.

The **deep antebrachial artery** (Figs. 102, 103) is a caudal branch of the median. Follow it deep to the flexor carpi radialis and transect the humeral head of the deep digital flexor which covers the vessel. The deep antebrachial artery supplies the flexor carpi radialis, the deep digital flexor, the flexor carpi ulnaris and the superficial digital flexor.

The **radial artery** (Figs. 102, 103) arises from the median artery in the middle of the forearm. It follows the medial border of the radius. At the carpus it divides into palmar and dorsal carpal branches that supply the deep vessels of the forepaw. These need not be dissected.

The **median artery** (Fig. 102) is the principal blood supply to the paw. It is accompanied by the median nerve along the humeral head of the deep digital flexor. The median artery passes through the carpal canal with the deep digital flexor tendon. Transect the flexor retinaculum. Follow this artery through the carpal canal to the proximal end of the metacarpus, where it forms the superficial palmar arch with a branch of the caudal interosseus artery. This arch gives rise to the palmar common digital arteries to the palmar surface of the forepaw (these need not be dissected).

NERVES OF THE FOREARM AND PAW

The skin of the antebrachium is innervated by three nerves: cranial and lateral surface by the radial nerve; caudal surface by the ulnar nerve; medial surface by the musculocutaneous nerve.

1. The **radial nerve** (Fig. 106) supplies the extensors of the elbow,

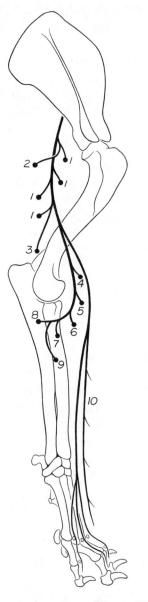

1. *Triceps brachii*
2. *Tensor fasciae antebrachii*
3. *Anconeus*
4. *Extensor carpi radialis*
5. *Supinator*
6. *Common digital extensor*
7. *Lateral digital extensor*
8. *Ulnaris lateralis*
9. *Abductor pollicis longus*
10. *Skin of cranial and lateral antebrachium and dorsal paw*

Figure 106. *Distribution of radial nerve, lateral view.*

carpus and digital joints. Reflect the lateral head of the triceps. Observe the radial nerve near the elbow where it divides into superficial and deep branches. Transect the extensor carpi radialis and reflect the proximal end. The **deep branch** of the radial nerve crosses the medial surface of the extensor carpi radialis in its course into the forearm with the brachialis muscle. Transect the supinator which lies over the radial nerve. The radial nerve innervates the extensor carpi radialis, the common digital extensor, the supinator, the lateral digital extensor, the abductor pollicis longus and the ulnaris lateralis.

The **superficial branch** divides into medial and lateral branches. The small **medial branch** follows the medial ramus of the cranial superficial antebrachial artery and continues distally in the forearm on the medial side of the cephalic vein. The **lateral branch** becomes associated with the lateral side of the cephalic vein and enters the forearm with the cranial superficial antebrachial artery. These branches continue to the paw on either side of the accessory cephalic vein. These medial and lateral branches are sensory to the skin on the cranial and lateral surface of the forearm and the dorsal surface of the carpus, metacarpus and digits. Do not dissect these nerves.

2. The **median nerve** (Figs. 103, 104) runs distally into the antebrachium with the median artery. It innervates the pronator teres, the pronator quadratus, the flexor carpi radialis, the superficial digital flexor and the radial head and parts of the humeral and ulnar heads of the deep digital flexor.

Reflect the flexor carpi radialis and the humeral head of the deep digital flexor to observe the course of the nerve. The median nerve passes through the carpal canal and branches to supply sensory innervation to the palmar surface of the forepaw. Note the course of the nerve through the carpal canal but do not dissect its terminal branches in the metacarpus.

3. The **ulnar nerve** (Fig. 105) diverges caudally from the median nerve at the distal third of the arm. At this point the caudal cutaneous antebrachial nerve arises from the ulnar. The ulnar nerve enters the antebrachial muscles caudal to the distal end of the humerus and is distributed to the flexor carpi ulnaris, parts of the ulnar and humeral heads of the deep digital flexor and the intrinsic muscles of the forepaw. In the middle of the antebrachium the dorsal branch of the ulnar nerve arises and becomes subcutaneous on the lateral surface. It is distributed to the lateral surface of the metacarpus and the fifth digit.

Transect and reflect the ulnar and humeral heads of the flexor carpi ulnaris. The ulnar nerve lies on the caudal surface of the deep digital flexor deep to the humeral head of the flexor carpi ulnaris.

Trace the palmar branch of the ulnar nerve into the forepaw. The nerve lies on the deep surface of the flexor carpi ulnaris just above the carpus. It then enters the lateral side of the carpal canal where it divides into a superficial and a deep branch. Reflect the flexor carpi ulnaris and flexor retinaculum to expose the nerve in the carpal canal. In the metacarpus the superficial and deep branches further divide to supply sensory innervation to the palmar surface of the forepaw and motor innervation to the intrinsic muscles of the forepaw. These branches will not be dissected.

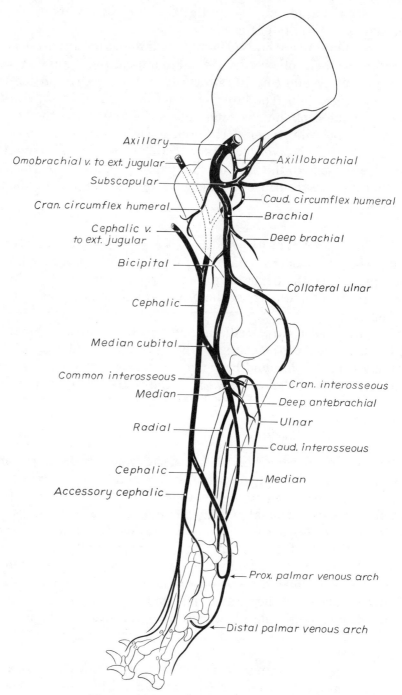

Figure 107.   *Veins of forelimb, medial view.*

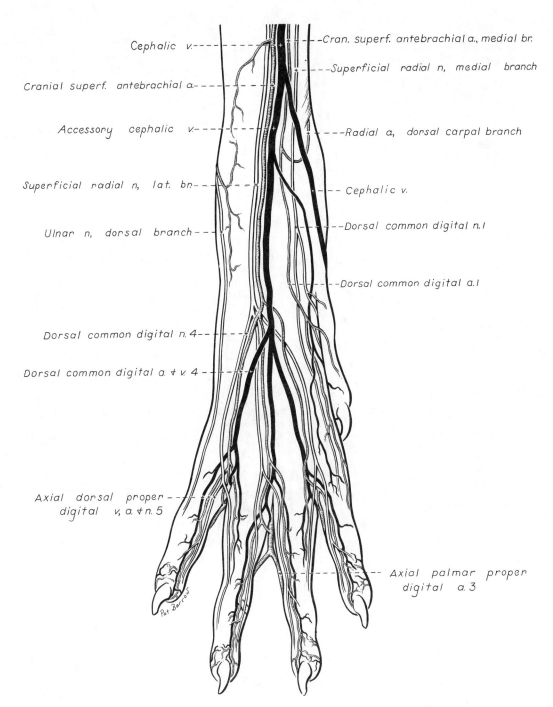

Cephalic v.

Cran. superf. antebrachial a., medial br.

Superficial radial n, medial branch

Cranial superf. antebrachial a.

Accessory cephalic v.

Radial a, dorsal carpal branch

Superficial radial n, lat. br.

Cephalic v.

Ulnar n, dorsal branch

Dorsal common digital n. 1

Dorsal common digital a. 1

Dorsal common digital n. 4

Dorsal common digital a. & v. 4

Axial dorsal proper
digital v, a. & n. 5

Axial palmar proper
digital a. 3

Pat Barrow

Figure 108. *Superficial structures of forepaw, dorsal view.*

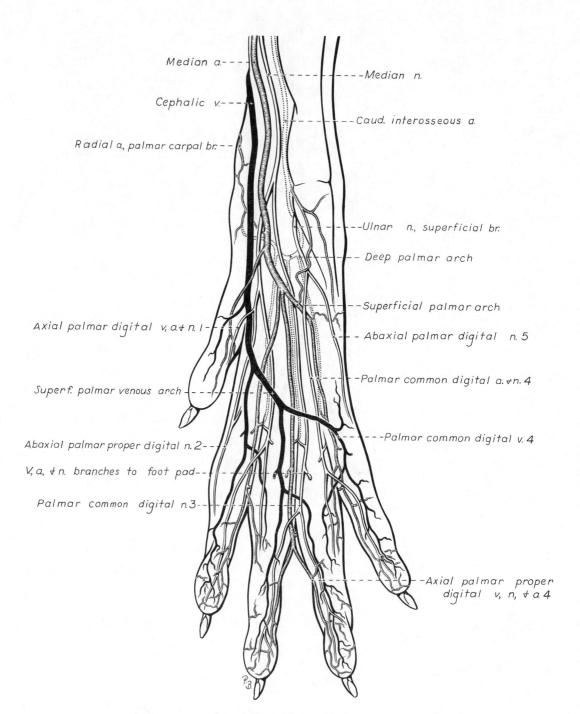

Figure 109. *Superficial structures of forepaw, palmar view.*

AREA	ARTERIAL SUPPLY	NERVE SUPPLY
Lateral Muscles of Scapula and Shoulder Stabilizers, flexors and extensors of shoulder Supraspinatus, infraspinatus	Superficial cervical	Suprascapular
Caudal Muscles of Scapula and Shoulder Flexors of Shoulder Deltoideus, Teres major, Teres minor	Subscapular	Axillary
Cranial Muscles of Arm Flexors of elbow, extensor of shoulder Biceps brachii (Brachialis)	Superficial cervical, Axillary, Brachial	Musculocutaneous
Caudal Muscles of Arm Extensor of elbow Triceps brachii	Axillary, Brachial	Radial
Cranial Muscles of Forearm Carpal extensors Digital extensors	Brachial: cranial interosseous	Radial
Caudal Muscles of Forearm Carpal flexors Digital flexors	Brachial: common interosseous Median: deep antebrachial	Median and ulnar
Dorsal Surface of Paw	Superficial brachial Dorsal carpal rete	Radial
Palmar Surface of Paw	Median Caudal interosseus	Median and Ulnar

DORSAL PAW
ARTERIES

(Superficial)	Superficial brachial a.
	    Cranial superficial antebrachial;
	    Medial br. of cranial superficial
	    antebrachial
		Dorsal
		common
		digital a.
			Axial or
			Abaxial

(Deep)	Dorsal carpal rete
	    Radial a., dorsal carpal br.
	    Caudal interosseous a.,
		Dorsal
		metacarpal a.
			dorsal
			digital a.

NERVES

(Superficial only)	Superficial br. radial n., medial and lateral br.	Dorsal common digital n.	Axial or Abaxial dorsal digital n.

PALMAR PAW
ARTERIES

(Superficial)	Superficial palmar arch,
	    Median a.;
	    Br. caudal interosseous a.
		Palmar
		common
		digital a.
			Axial or
			Abaxial
			palmar
			digital a.

(Deep)	Deep palmar arch
	    Caudal interosseous a.
	    Palmar carpal br., radial a.
		Palmar
		metacarpal a.	

NERVES

(Superficial)	Median n. Ulnar n., superficial br.	Palmar common digital nn.	Axial or Abaxial palmar digital n.
(Deep)	Ulnar n., deep br.	Palmar metacarpal nn. 1–4	

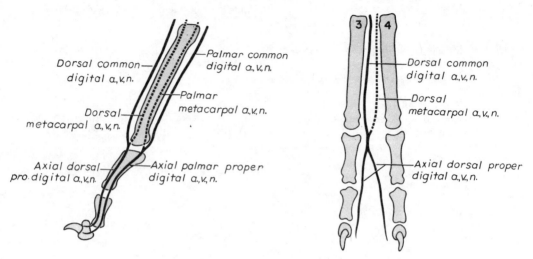

Figure 110a.   *Schema of blood supply and innervation of digits.*

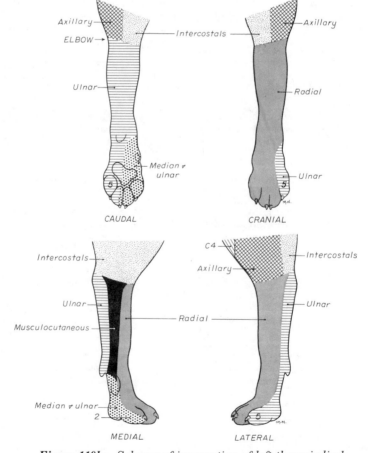

Figure 110b.   *Schema of innervation of left thoracic limb.*

THE ABDOMEN, PELVIS AND PELVIC LIMB

Reflect the skin from the right abdominal wall leaving the mammary papillae in the female, the prepuce in the male and the cutaneous trunci. Extend a perpendicular incision from the ventral midline to the middle of the medial surface of the right thigh and thence to its cranial border. Continue this incision dorsally along the cranial edge of the thigh past the crest of the ilium to the middorsal line. Starting on the medial surface of the thigh, reflect or remove the skin of the right side of the abdomen.

VESSELS AND NERVES OF THE VENTRAL AND LATERAL PARTS OF THE ABDOMINAL WALL

The arteries which supply the superficial part of the ventral abdominal wall are branches of the superficial epigastric arteries.

The subcutaneous tissue of the ventral abdominal wall contains the abdominal and inguinal mammae and the vessels and nerves which supply them. In the female the cranial superficial epigastric vessels are seen subcutaneously near the cranial abdominal papilla. Make a midventral incision through the fat-filled tissue. By blunt dissection separate the right row of mammae from the fascia and turn them laterally.

Dissect the **external pudendal artery** (Figs. 111, 133) which emerges from the superficial inguinal ring. Its origin from the pudendoepigastric trunk, a branch from the deep femoral artery, will be seen later. The external pudendal artery courses caudoventrally to the cranial border of the gracilis. The **caudal superficial epigastric artery** arises from the external pudendal dorsal to the superficial inguinal lymph node. The caudal superficial epigastric artery (Fig. 111) runs cranially to the deep surface of the inguinal mamma and supplies the mammary branches. The artery continues its deep course to supply the caudal abdominal mamma and anastomose with branches of the cranial superficial epigastric artery. The external pudendal continues caudoventrally to supply the vulva or scrotum.

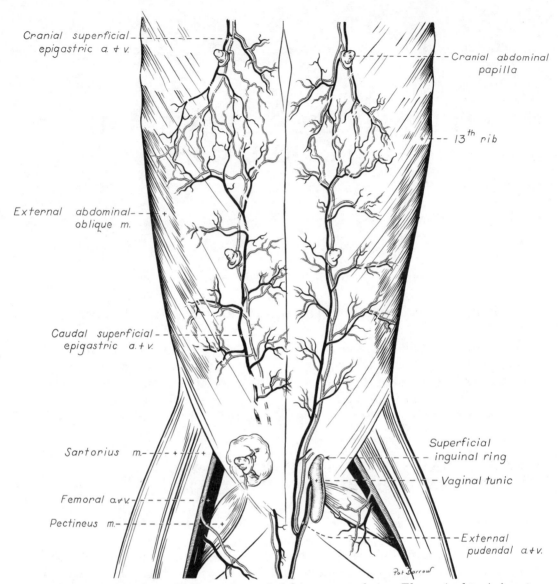

Figure 111. *Superficial structures of abdomen, ventral view. The vaginal tunic is not exposed on the right side of the specimen.*

Expose the **superficial inguinal lymph nodes** which lie adjacent to the caudal superficial epigastric vessels and cranial to the pudendal vessels. The afferent lymphatics of these nodes drain the mammae, the prepuce, the scrotum and the ventral abdominal wall as far cranially as the umbilicus.

Reflect the superficial fascia from the lateral abdominal wall. Emerging from the dorsolateral abdominal wall, caudal to the last rib, are superficial branches of the **cranial abdominal artery.** The latter arises from the phrenicoabdominal artery, a branch of the aorta, and perforates the abdominal musculature to reach the skin.

The **cutaneous nerves** of the abdomen differ somewhat from those of the thorax. The lateral cutaneous branches from the last five thoracic nerves do not follow the convexity of the costal arch but run in a caudo-ventral direction and supply most of the ventral and ventrolateral parts of the abdominal wall (Fig. 112). The cutaneous branches of the first three lumbar nerves perforate the midlateral part of the abdominal wall and as small nerves run caudoventrally. They supply the skin of the caudolateral abdominal wall and the thigh in the region of the stifle. Do not dissect these cutaneous nerves. Cranial to the ventral iliac spine the **lateral cutaneous femoral nerve** and the **deep circumflex iliac artery** and **vein** perforate the internal abdominal oblique and appear superficially. The nerve arises from the **fourth lumbar nerve** and is cutaneous to the cranial and lateral surfaces of the thigh. Dissect these vessels and trace the nerve as far as the present skin reflection will allow.

Complete the transection of the origin of the right external abdominal oblique which was started when the rib cage was reflected. Transect all but the last centimeter of the internal abdominal oblique 2 cm. dorsal

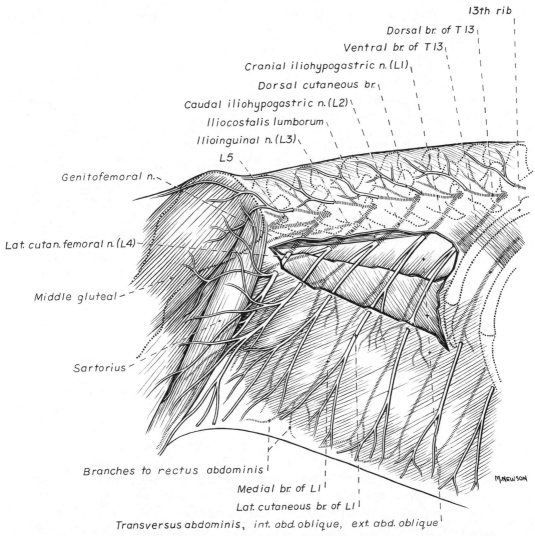

Figure 112. *Lateral view of the first four lumbar nerves.*

to its aponeurosis. Separate the oblique muscles from the underlying transverse muscle to expose the ventral branches of the last five thoracic nerves and the first four lumbar nerves. These are parallel to each other and supply the ventral and lateral parts of the thoracic and abdominal wall. The first four lumbar nerves form the cranial iliohypogastric, caudal iliohypogastric, ilioinguinal and lateral cutaneous femoral nerves respectively.

The cranial and caudal iliohypogastric and ilioinguinal nerves (Fig. 112) pass through the aponeurosis of origin of the transversus abdominis. Each has a medial branch which descends between the transversus abdominis and the internal abdominal oblique to the rectus abdominis. The medial branches supply these muscles, the underlying peritoneum and the skin of the ventral abdominal wall. The lateral branches of these nerves perforate the internal abdominal oblique and descend between the oblique muscles. They may be seen on the deep surface of the external abdominal oblique. Each lateral branch supplies these muscles, perforates the external abdominal oblique and terminates subcutaneously as the lateral cutaneous branch to the abdominal wall in that region.

INGUINAL STRUCTURES

Dissect the structures in the male which pass through the inguinal canal and the superficial inguinal ring (Fig. 113).

MALE:

The **external pudendal artery** and **vein** leave the superficial inguinal ring caudal to the structures which supply the testis. Their branches have been dissected.

The **genitofemoral nerve** arises from the ventral branches of the third and fourth lumbar nerves. It is bound by fascia to the external pudendal vein medial to the spermatic cord. It innervates the skin covering the inguinal region of both sexes and that of the scrotum and part of the penis in the male.

The **spermatic fascia** surrounds the structures emerging from the superficial inguinal ring. Reflect it from the underlying parts.

The **cremaster muscle** is surrounded by this spermatic fascia as it courses along the caudal part of the vaginal tunic. This tunic was exposed upon the reflection of the spermatic fascia and can be seen extending from its emergence through the superficial inguinal ring to the testis. The cremaster muscle arises from the caudal free border of the internal abdominal oblique and attaches to the vaginal tunic near the testis.

The **vaginal tunic** (Figs. 114–116) is a diverticulum of the peritoneum present in both sexes. In the male it envelops the structures of the spermatic cord. It consists of the following parts in the male:

The **parietal vaginal tunic,** the outer layer of the vaginal tunic, extends from the deep inguinal ring to the bottom of the scrotum.

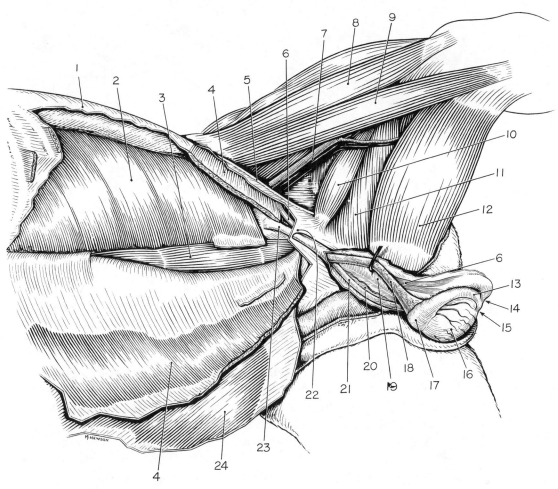

Figure 113. Abdominal muscles and inguinal region of the male, deep dissection.

1. Thoracolumbar fascia
2. Transversus abdominis
3. Rectus abdominis
4. Internal abdominal oblique (transected and reflected)
5. Inguinal ligament
6. Cremaster muscle
7. Femoral artery and vein
8. Cranial part of sartorius
9. Caudal part of sartorius
10. Pectineus
11. Adductor
12. Gracilis
13. Tail of epididymis
14. Ligament of tail of epididymis
15. Proper ligament of testis
16. Testis in visceral vaginal tunic
17. Head of epididymis
18. Testicular artery and vein in visceral vaginal tunic (mesorchium)
19. Mesorchium
20. Mesoductus deferens
21. Ductus deferens in visceral vaginal tunic
22. Superficial inguinal ring
23. Parietal vaginal tunic
24. External abdominal oblique (reflected)

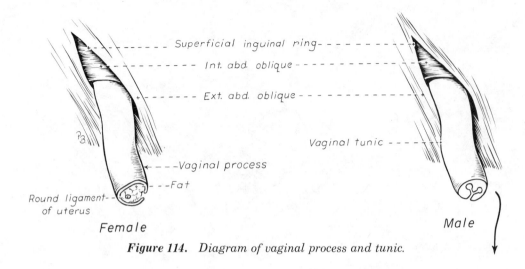

Figure 114. *Diagram of vaginal process and tunic.*

Open this tunic along the most ventral part of the testis and along the cranial border of the vaginal tunic to the superficial inguinal ring to expose the visceral vaginal tunic.

The **visceral vaginal tunic** is closely fused to the testis and epididymis, and surrounds the ductus deferens and vessels and nerves which arise or terminate in the testis and epididymis. The **mesorchium** is the part of the visceral vaginal tunic that contains the vessels and nerves of the testis. The **mesoductus deferens** is the part of the visceral vaginal tunic that attaches to the ductus deferens.

The **spermatic cord** (Fig. 115) is carried through the inguinal canal by the descent of the testis and is composed of two distinct parts:

The **ductus deferens** carries the spermatozoa from the epididymis to the urethra. It arises from the tail of the epididymis at the caudal end of the testis and is attached to the mesorchium by the **mesoductus deferens.** The small **deferent artery** and **vein** accompany the deferent duct.

The **testicular artery** and **vein,** as well as the testicular lymph vessels and the testicular plexus of autonomic nerves, are closely

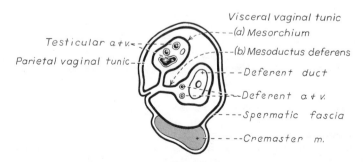

Figure 115. *Vaginal tunic of male, transverse section.*

associated with each other. These vessels and nerves are covered by a fold of visceral vaginal tunic, the **mesorchium,** which is continuous with the parietal vaginal tunic. The artery is tortuous, and woven around it are the nerve plexus and the venous plexus. The venous plexus is the **pampiniform plexus.** The nerve plexus is autonomic and contains postganglionic sympathetic axons which arise from the third to fifth lumbar sympathetic ganglia.

The **testis** and **epididymis** (Fig. 113) are intimately covered by the visceral vaginal tunic. At the caudal extremity of the epididymis the visceral peritoneum leaves the epididymis at an acute angle and becomes the parietal layer. Thus there is a small circumscribed area on the epididymis not covered by peritoneum. The connective tissue which attaches the epididymis to the vaginal tunic and spermatic fascia at this point is the **ligament of the tail of the epididymis.** Reflect the skin of the scrotum caudally to observe this.

The epididymis (Fig. 113) lies more on the lateral side of the testis than on its dorsal border. For descriptive purposes it is divided into a cranial extremity or **head,** where the epididymis communicates with the testis; a middle part or **body;** and a caudal extremity or **tail,** which is continuous with the ductus deferens. The tail is attached to the testis by the **proper ligament of the testis.**

Lay the previously reflected skin back over the inguinal region and examine the scrotum. The **scrotum** is a pouch divided by an external raphe and an internal median septum into two cavities, each of which is occupied by a testis, an epididymis and the distal part of the spermatic cord.

FEMALE:

In the female, locate the external pudendal blood vessels and the genitofemoral nerve emerging from the superficial inguinal ring. The **vaginal process** is accompanied by the **round ligament of the uterus** whose origin within the abdomen from the mesometrium will be seen later. These two structures, enclosed in fascia and surrounded by fat, may extend caudally as far as the vulva.

THE INGUINAL CANAL

The **inguinal canal** (Fig. 116) is a short fissure filled with connective tissue between the abdominal muscles. It extends between the deep and superficial inguinal rings. It is bounded laterally by the aponeurosis of the external abdominal oblique, cranially by the caudal border of the internal abdominal oblique, caudally by the inguinal ligament and medially in part by the superficial surface of the rectus abdominis, the transversus abdominis and the internal abdominal oblique. The vaginal tunic and spermatic cord pass obliquely caudoventrally through the canal. Notice as many of these boundaries as possible before opening the abdomen.

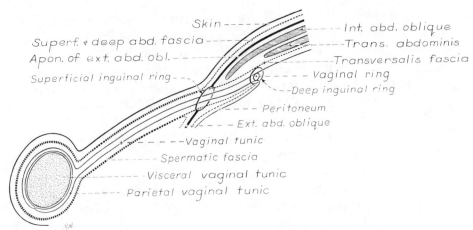

Figure 116. Schema of vaginal tunic and its coverings.

ABDOMINAL AND PERITONEAL CAVITIES

The **transversalis fascia** reinforces the peritoneum and attaches it to the abdominal muscles and diaphragm. Make a paramedian incision through the abdominal wall on each side dorsal to the rectus abdominis from the costal arch to the level of the inguinal canal. Connect the cranial ends of these incisions and reflect the ventral abdominal wall. Observe the following structures:

The **falciform ligament** is a fold of peritoneum which passes from the umbilicus to the diaphragm. It is also attached to the liver between the left and right medial lobes. In obese specimens a large accumulation of fat is found in this remnant of the ventral mesentery. In young animals the **round ligament of the liver** may still be visible in the free border of the falciform ligament. Caudal to the umbilicus the fold of peritoneum is the **median ligament of the bladder.**

The **vaginal ring** is formed by the parietal peritoneum as it leaves the abdomen and enters the inguinal canal to form the vaginal tunic. It marks the position of the **deep inguinal ring,** which is formed by the reflection of the transversalis fascia outside the vaginal ring. A deposit of fat is usually present in the transversalis fascia around the vaginal ring.

The **ductus deferens** is attached to the abdominal and pelvic walls by a fold of peritoneum, the mesoductus deferens. At the vaginal ring this fold joins the mesorchium, which contains the testicular artery and vein and the testicular nerve plexuses.

The **caudal epigastric artery** and **vein** course cranially on the deep face of the caudal part of the rectus abdominis. The origin of the artery from the pudendo-epigastric trunk of the deep femoral artery will be dissected later.

ABDOMINAL VISCERA

The **greater omentum** (Figs. 117–119) is the first structure seen after reflecting the abdominal wall. It is a double-walled peritoneal sac, lace-

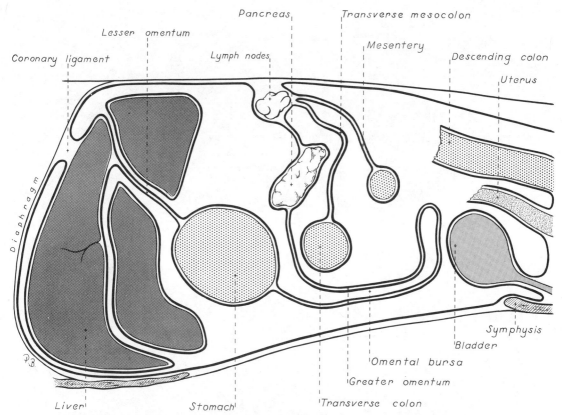

Figure 117. *Diagram of peritoneal reflections, sagittal section.*

like in appearance, with depositions of fat along the vessels. The greater omentum covers the jejunum and ileum, leaving the descending colon exposed on the left, the bladder exposed caudally and the descending duodenum exposed on the right. Reflect the omentum and separate its superficial and deep walls to demonstrate its cavity, the **omental bursa.**

Three organs in the abdomen that are capable of considerable variation in size are the stomach, the urinary bladder and the uterus. If one or more of these is distended the relations of other organs will be altered.

The **urinary bladder** (Fig. 121), when empty, is contracted and lies on the floor of the pelvic inlet. When distended, it lies on the floor of the abdomen and conforms in shape to the caudal part of the abdominal cavity since it displaces all freely movable viscera. It frequently reaches a transverse plane through the umbilicus.

The non-pregnant **uterus** (Fig. 120) is remarkably small even in a bitch that has had several litters. The uterus consists of a **cervix,** a **body** and two **horns.** The gravid uterus lies on the floor of the abdomen during the second month or last half of pregnancy. As the uterus enlarges, the middle parts of the horns gravitate cranially and ventrally and come to lie medial to the costal arches; thus the uterus bends on itself as the ovarian and vaginal ends move very little during enlargement.

The **broad ligaments** of the uterus attach the ovary, the uterine tube and the uterus to the lateral part of the sublumbar region. A fold arises

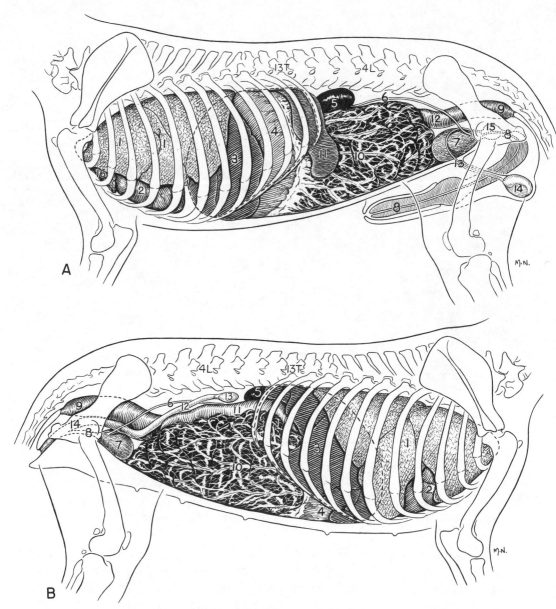

Figure 118. *Viscera of the dog*

A Viscera of male dog, left lateral view.
1. Left lung
2. Heart
3. Liver
4. Stomach
5. Left kidney
6. Ureter
7. Bladder
8. Urethra
9. Rectum
10. Greater omentum covering small intestine
11. Spleen
12. Descending Colon
13. Ductus deferens
14. Left testis
15. Prostate
16. Thymus

B Viscera of female dog, right lateral view.
1. Right lung
2. Heart
3. Liver
4. Stomach
5. Right kidney
6. Ureter
7. Bladder
8. Urethra
9. Rectum
10. Greater omentum covering small intestine
11. Descending duodenum
12. Right uterine horn
13. Right ovary
14. Vagina

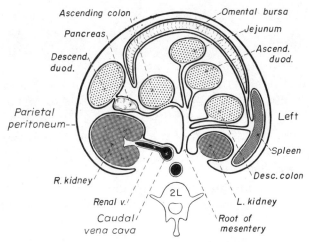

Figure 119. Abdominal mesenteries.

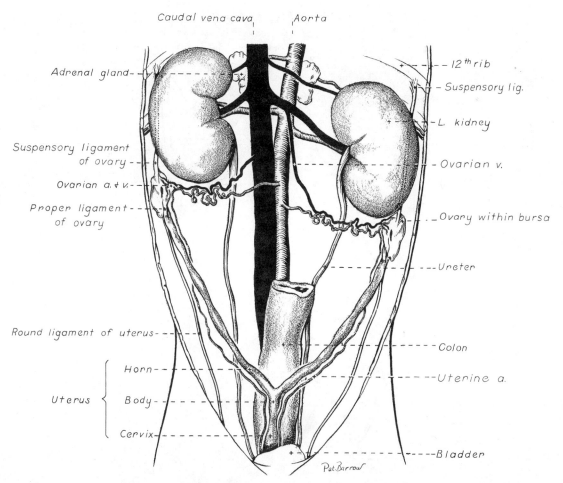

Figure 120. Schema of abdominal part of urogenital system.

from the lateral peritoneal layer of each broad ligament. In the free border of this peritoneal fold is the **round ligament** of the uterus. The round ligament arises near the ovary and runs through the inguinal canal along with the vaginal tunic.

The **spleen** (Fig. 121) lies in the greater omentum to the left of the median plane along the greater curvature of the stomach. Its position, shape and degree of distension are variable. Its lateral surface lies against the parietal peritoneum of the left lateral abdominal wall and the liver. Its caudal part may reach to a transverse plane through the mid-lumbar region. Its cranial limit is usually marked by a plane passing between the twelfth and thirteenth thoracic vertebrae. It may reach the floor of the abdomen. The part of the greater omentum which attaches the spleen to the stomach is the **gastrosplenic ligament.**

The **diaphragm** (Fig. 122), the muscular partition between the thoracic and the abdominal cavities, is a muscle of inspiration. It has an extensive muscular periphery and a small, V-shaped **tendinous center.**

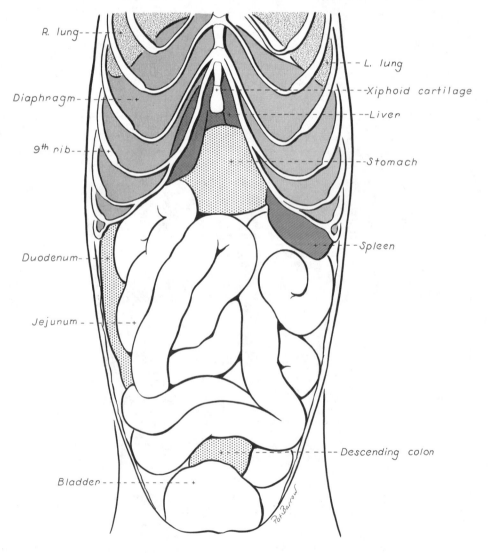

Figure 121.   *Schematic ventral view of abdominal viscera.*

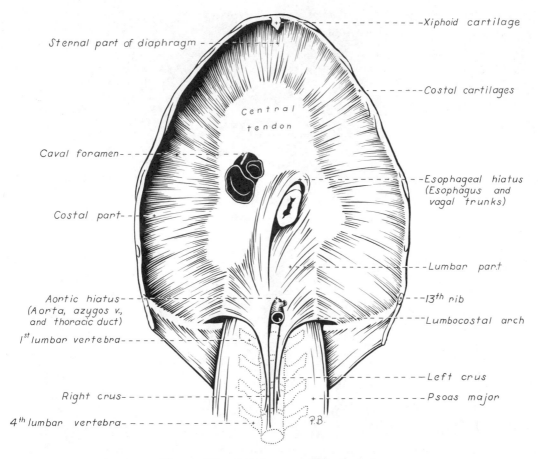

Xiphoid cartilage

Sternal part of diaphragm

Costal cartilages

Central tendon

Caval foramen

Esophageal hiatus
(Esophagus and
vagal trunks)

Costal part

Lumbar part

Aortic hiatus
(Aorta, azygos v,
and thoracic duct)

13th rib

Lumbocostal arch

1st lumbar vertebra

Left crus

Right crus

Psoas major

4th lumbar vertebra

P.B.

Figure 122. Diaphragm, abdominal view.

The muscular part of the diaphragm may be divided into three parts according to its attachments: lumbar, costal and sternal. The right crus and left crus form the **lumbar part** and attach to the bodies of the third and fourth lumbar vertebrae. The right crus is larger than the left. The **costal part** of the diaphragm arises from the medial surfaces of the eighth to thirteenth ribs. It interdigitates with the transversus abdominis muscle. The **sternal part** arises from the dorsal surface of the sternum cranial to the xiphoid cartilage. The extensions of the V-shaped tendinous center run dorsally between the lumbar and costal parts on each side. The caudal mediastinum may be severed to expose the tendinous part of the muscle.

The **aortic hiatus** is a passageway between the crura for the aorta, the azygos vein and the thoracic duct. The more centrally located **esophageal hiatus** transmits the esophagus, vagal nerve trunks and esophageal vessels. The **caval foramen** is located at the junction of the tendinous and muscular parts of the right side of the diaphragm. The caudal vena cava passes through it.

The **liver** (Fig. 123) has six lobes, and its parietal surface conforms to the abdominal surface of the diaphragm. The visceral surface of the liver is related on the left to the stomach and sometimes to the spleen; on the

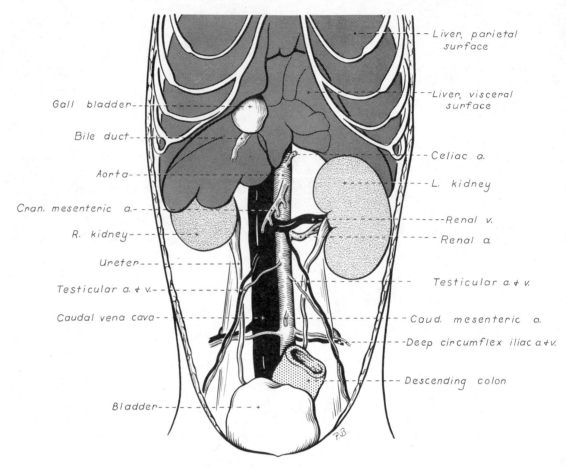

Gall bladder

Bile duct

Aorta

Cran. mesenteric a.

R. kidney

Ureter

Testicular a. & v.

Caudal vena cava

Bladder

Liver, parietal surface

Liver, visceral surface

Celiac a.

L. kidney

Renal v.

Renal a.

Testicular a. & v.

Caud. mesenteric a.

Deep circumflex iliac a. & v.

Descending colon

Figure 123.   Abdominal viscera, after removal of stomach, pancreas, spleen and intestine.

right to the pancreas, right kidney and duodenum; ventrally to the greater omentum and through this to the intestinal coils. Its most caudal part covers the cranial extremity of the right kidney and reaches a transverse plane through the thirteenth thoracic vertebra. The liver rarely projects caudal to the costal arch. It undergoes slight longitudinal movement with each respiration.

The **right medial lobe** of the liver contains a fossa for the gall bladder. The **right lateral lobe,** which is smaller, is located next to the caudate lobe which embraces the cranial end of the right kidney. The **quadrate lobe** is narrow and is located between the right and left medial lobes. It forms the left boundary of the fossa of the gall bladder. The **left medial lobe** is separated by a fissure from the right medial and quadrate lobes. The **left lateral lobe** is separated by a fissure from the left medial lobe. The free margin of the left lateral lobe is frequently notched. The visceral surface of the left lateral lobe is concave where it contacts the stomach. The **caudate lobe** is indistinctly separated from the central mass of the liver. It lies transversely, but is mainly to the right of and dorsal to the main bulk of the organ. It is constricted in its middle; its extremities are in the form of two processes. The **caudate process** caps the cranial end of the right kidney and thus contains the deep **renal impression.** The **papillary process** can be seen through the lesser omentum if the liver is tipped forward. It lies in the lesser curvature of the stomach.

The **stomach** (Figs. 118, 121, 124) is divided into parts which blend imperceptibly with one another. The **cardiac part** is the smallest part of the stomach and is situated nearest the esophagus. The **fundus** is dome-shaped and lies to the left of and dorsal to the cardia. The **body** of the stomach is the large middle portion. It extends from the fundus on the left to the pyloric part on the right. The **pyloric part** is constricted and joins the duodenum at a sphincter, the **pylorus.**

The stomach is bent so that its **greater curvature** faces mainly to the left and its **lesser curvature** mainly to the right; its parietal surface faces ventrally toward the liver and its visceral surface dorsally faces the intestinal mass. Its position changes depending on its fullness.

The **empty stomach** is completely hidden from palpation and observation by the liver and diaphragm cranioventrally and the intestinal mass caudally. It lies to the left of the median plane. The empty stomach is cranial to the costal arch and sharply curved so that it is more V-shaped than C-shaped. The greater curvature faces ventrally, caudally and to the left. This curvature lies above and to the left of the mass of the

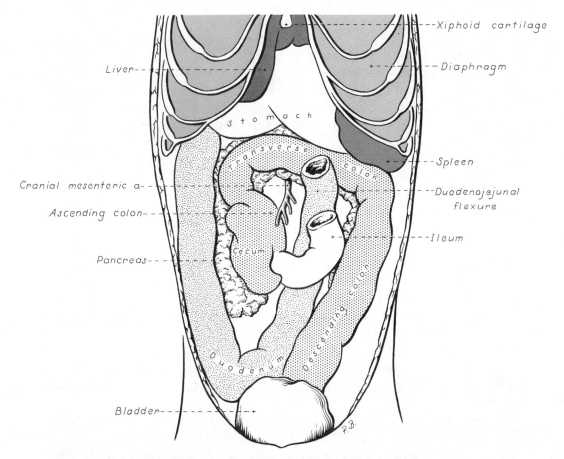

Figure 124. Abdominal viscera, jejunum removed.

intestines. The lesser curvature is strongly curved around the papillary process of the liver and faces craniodorsally and to the right.

The **full stomach** lies in contact with the ventral abdominal wall and protrudes beyond the costal arches. It displaces the intestinal mass. Open the stomach along its parietal surface, remove the contents and observe the longitudinal folds of mucosa.

The **duodenum** (Fig. 124) is the most fixed part of the small intestine. It begins at the pylorus to the right of the median plane. After a short dorsocranial course it turns as the **cranial duodenal flexure.** It continues caudally on the right as the **descending part,** where it is in contact with parietal peritoneum. Farther caudally the duodenum turns, forming the **caudal duodenal flexure,** and continues cranially as the **ascending part.** The ascending part lies to the left of the root of the mesentery, where it forms the **duodenojejunal flexure.**

The **jejunum** and the **ileum** form the coils of the small intestine (Fig. 121) which occupy the ventrocaudal part of the abdominal cavity. They receive their nutrition from the cranial mesenteric artery which is in the root of the mesentery. The root of the mesentery attaches the jejunum and ileum to the dorsal body wall. The **mesenteric lymph nodes** lie along the vessels in the mesentery. The jejunum begins at the left of the root of the mesentery and is the longest portion of the small intestine. The ileum is the terminal portion of the small intestine. It passes from the left to the right side in a transverse plane through the midlumbar region caudal to the root of the mesentery, and joins the colon on the right of the median plane at the **ileocolic orifice.**

The **cecum,** a part of the large intestine, is an S-shaped, blind tube located to the right of the median plane at the junction of the ileum and colon. It is ventral to the caudal extremity of the right kidney, dorsal to the small intestines and medial to the descending part of the duodenum. The cecum communicates with the ascending colon at the **cecocolic orifice.** Open the intestine at the ileocolic junction and observe the orifices.

The **colon** (Fig. 124) is divided into a short **ascending colon** which lies on the right of the root of the mesentery, a **transverse colon** which lies cranial to the root of the mesentery and a long **descending colon** which lies at its beginning on the left of the root of the mesentery. The bend between the ascending and transverse colons is known as the **right colic flexure** and that between the transverse and descending colons is known as the **left colic flexure.** The descending colon terminates at a transverse plane through the pelvic inlet. It is continued by the **rectum.**

The **pancreas** (Fig. 125) is lobulated and is composed of a body and two lobes. The **body** lies at the pylorus. The **right lobe** lies ventral to the right kidney and dorsomedial to the descending part of the duodenum. It is enclosed by the mesoduodenum. The **left lobe** of the pancreas lies between the peritoneal layers which form the dorsal wall of the omental bursa.

The pancreatic duct system (Fig. 125) is variable. Most dogs have two ducts; these open separately in the duodenum but communicate in the gland. The **pancreatic duct** is the smaller of the two ducts and is some-

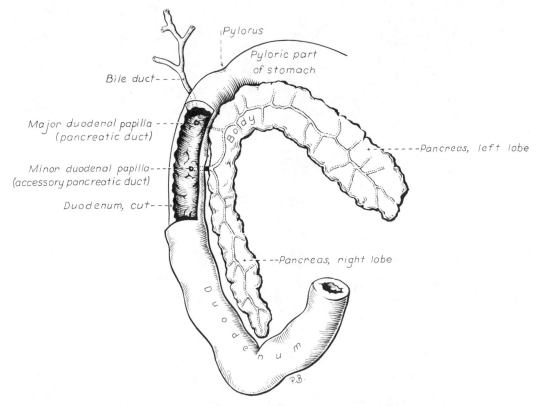

Figure 125. Pancreas with pancreatic and bile ducts.

times absent. It opens close to the bile duct on the **major duodenal papilla.** Make an incision through the free border of the descending part of the duodenum. Scrape away the mucosa with the scalpel handle and identify the major duodenal papilla. The **accessory pancreatic duct** opens into the duodenum on the **minor duodenal papilla** caudal to the major papilla. Locate the accessory duct by blunt dissection in the mesoduodenum between the right lobe of the pancreas and the descending duodenum.

The **adrenal glands** (Fig. 126) are light-colored and are located at the cranial aspect of each kidney. Each gland is crossed by the phrenicoabdominal vein, which leaves a deep groove on its ventral surface.

The **right adrenal gland** lies between the caudal vena cava and the liver ventrally and the sublumbar muscles dorsally. Expose the gland by dissection between the caudal vena cava and the kidney cranial to the renal vein.

The **left adrenal gland** lies between the aorta and the left kidney.

The **kidneys** (Figs. 126, 127) are dark brown, are partly surrounded by fat and are covered on their ventral surface by peritoneum. The lateral border is strongly convex and the medial, nearly straight. At the middle of the medial border is an indentation, the **hilus** of the kidney, where the renal vessels and nerves and the ureter communicate with the organ.

The **right kidney** lies opposite the first three lumbar vertebrae. It is

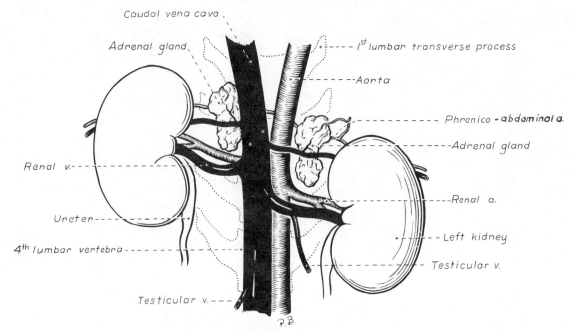

Figure 126.   *Kidneys and adrenal glands.*

farther cranial than the left kidney by the length of half a kidney. The right kidney is more extensively related to the liver than to any other organ. Its cranial third is embedded in the caudate process of the caudate lobe of the liver. The remaining ventral surface is related to the descending duodenum, the right lobe of the pancreas, the cecum and the ascending colon. The caudal vena cava is on the medial border of the right kidney.

The **left kidney** lies opposite the second, third and fourth lumbar vertebrae. It is related ventrally to the descending colon and the small intestine. The spleen is related to the cranial extremity of the gland. The medial border is close to the aorta.

The expanded part of the **ureter** within the kidney is the **renal pelvis.** The ureter opens into the dorsal part of the neck of the urinary bladder. Throughout this course it is enveloped by a fold of peritoneum from the dorsal body wall. Follow the course of the ureter. The **renal sinus** is the fat-filled space that contains the renal vessels and surrounds the renal pelvis.

Free the left kidney from its covering peritoneum and fascia. Do not cut its vascular attachment. Make a longitudinal section of the left kidney from its lateral border to the hilus, dividing it into dorsal and ventral halves. Note the granular appearance of the peripheral portion of the renal parenchyma. This is the **renal cortex,** which contains primarily the renal corpuscles and convoluted portions of the tubules. The more centrally positioned parenchyma is the **medulla.** It is lighter in color and has a striated appearance due to numerous collecting ducts. The vessels that are apparent at the corticomedullary junction are the **arcuate branches** of the renal vessels. The longitudinal ridge projecting into the renal pelvis

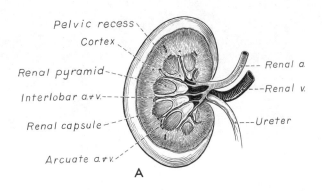

Pelvic recess

Cortex

Renal pyramid

Interlobar a.&v.

Renal capsule

Arcuate a.&v.

Renal a.

Renal v.

Ureter

A

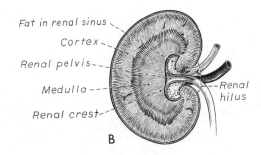

Fat in renal sinus

Cortex

Renal pelvis

Medulla

Renal crest

Renal hilus

B

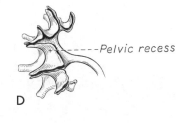

Pelvic recess

D

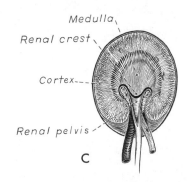

Medulla

Renal crest

Cortex

Renal pelvis

C

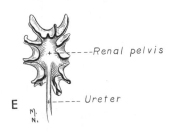

Renal pelvis

Ureter

E

Figure 127. *Details of structure of left kidney.*
A Sectioned in dorsal plane.
B Sectioned in mid-dorsal plane.
C Transverse section.
D Cast of renal pelvis, dorsal view.
E Cast of renal pelvis, medial view.

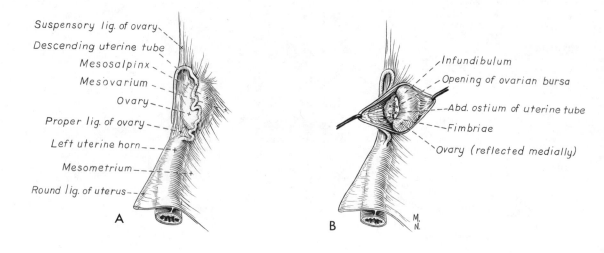

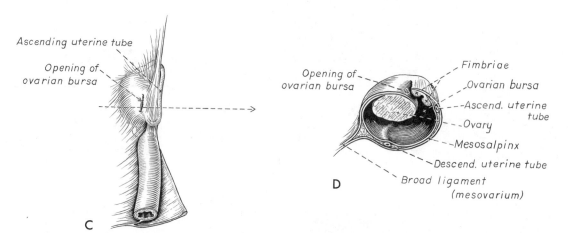

Figure 128.   *Relations of left ovary and ovarian bursa.*
A Dorsal view.
B Dorsal view, ovarian bursa opened.
C Ventral view.
D Section through ovary and ovarian bursa.

is the **renal crest,** through which collecting tubules of the kidney excrete urine into the renal pelvis. Make a second longitudinal section parallel to the first and note the **renal pyramids** formed by the medulla. The **pelvic recesses** of the renal pelvis project outward between the renal pyramids.

Free the right kidney from its peritoneum and fascia and make a transverse section through it. Note the renal cortex, medulla, crest and pelvis.

The **ovarian bursa** (Fig. 128) is a thin-walled peritoneal sac which encloses the ovary and is open to the peritoneal cavity via a slit-like orifice on the medial side. The **uterine tube** courses through the wall of the bursa. The infundibulum of the uterine tube is fimbriated.

Each ovary is attached by its **proper ligament** to the uterus and by its **suspensory ligament** to the transversalis fascia medial to the last rib. The right ovary lies cranial to the left and is dorsal to the descending duodenum. The left ovary is between the descending colon and the abdominal wall. Open the bursa and examine the ovary.

Find the uterine tube (Fig. 128) within the wall of the ovarian bursa. The ovarian end of the tube, the **infundibulum,** is located near the edge of the opening into the ovarian bursa. The infundibulum is a funnel-shaped dilatation of the lumen of the uterine tube which narrows into a minute opening, the **abdominal ostium.** The free edges of the infundibulum are fringed by numerous finger-like processes, the fimbriae, some of which project through the opening of the ovarian bursa. The uterine tube courses first cranially, then caudally, in the wall of the ovarian bursa. It terminates at the uterine horn.

The paired **broad ligament** of the uterus (Fig. 128) is the peritoneal fold which suspends the internal genitalia with the exception of the caudal part of the vagina, which is not covered by peritoneum. Each ligament is divided into three parts: The **mesometrium** arises from the lateral wall of the pelvis and the lateral part of the sublumbar region and attaches to the lateral part of the cranial end of the vagina, uterine cervix and uterine body, and the corresponding uterine horn. The **mesovarium,** a continuation of the mesometrium, is the cranial part of the broad ligament. It begins at a transverse plane through the cranial end of the uterine horn and attaches the ovary and the ligaments associated with the ovary to the lateral part of the sublumbar region. The **mesosalpinx** is the peritoneum which attaches the uterine tube to the mesovarium and forms with the mesovarium the walls of the ovarian bursa.

PERITONEUM

The peritoneum lines the abdominal, pelvic and scrotal cavities and reflects around their contained organs. The **parietal peritoneum** covers the inner surface of the walls of the abdominal, pelvic and scrotal cavities; the **visceral peritoneum** covers the organs of these cavities.

In the embryo the dorsal common mesentery is a double sheet of peritoneum which passes from the dorsal abdominal wall to the digestive tube. It serves as a route by which the nerves and vessels reach the organs. In the dog, the dorsal common mesentery persists as the mesoduodenum, mesentery, mesocolon and greater omentum.

The **mesoduodenum** originates from the dorsal abdominal wall and the root of the mesentery and extends to the duodenum. On the right side it passes to the descending duodenum and encloses the right lobe of the pancreas between its layers. Cranially it is continuous with the greater omentum across the ventral surface of the portal vein. Caudally the mesoduodenum passes from the root of the mesentery to the caudal flexure of the duodenum. On the left it is attached to the ascending duodenum and at the duodenojejunal flexure it is continuous with the mesentery of the jejunum. The ascending duodenum is attached to the mesocolon by the duodenocolic fold.

The **mesentery** (mesojejunoileum) attaches to the abdominal wall opposite the second lumbar vertebra by a short peritoneal attachment known as the **root of the mesentery.** Vessels and nerves pass in the mesentery to supply the large and small intestine. The peripheral border of the mesentery attaches to the jejunum and the ileum and is as long as the section of the intestine between them. At the ileocolic junction the mesentery is continuous with the ascending mesocolon.

The **ascending, transverse** and **descending mesocolons** connect the ascending, transverse and descending colons to the dorsal body wall. They are continuous with each other from right to left.

An **omentum** (epiploon) attaches the stomach to the body wall or other organs:

The **greater omentum** (see Figs. 117, 118) attaches the greater curvature of the stomach to the dorsal body wall. From the greater curvature of the stomach it extends caudally on the floor of the abdomen. It turns dorsally on itself near the pelvic inlet and returns dorsal to the stomach where it contains the left lobe of the pancreas between its peritoneal layers. It attaches to the dorsal abdominal wall. Caudal and to the left of the fundus of the stomach is the spleen, which lies largely in an outpocketing of the superficial part of the greater omentum.

The **lesser omentum** loosely spans the distance from the lesser curvature of the stomach to the porta of the liver. Between the liver and the cardia of the stomach it attaches for a short distance to the diaphragm. The papillary process of the liver is loosely enveloped by the lesser omentum. On the right a part of the lesser omentum, the **hepatoduodenal ligament,** attaches the liver to the duodenum. It contains the portal vein, the hepatic artery and the bile duct.

The **omental bursa** is formed by the omenta and the adjacent organs. It has an **epiploic foramen** opening into the main peritoneal cavity. This opening lies to the right of the median plane at the level of the cranial duodenal flexure, caudomedial to the caudate lobe of the liver. It is bounded dorsally by the caudal vena cava and ventrally by the portal vein. Find this foramen and insert a finger through it.

There are a few short peritoneal folds which serve more to fix organs in position than as channels for blood vessels.These are called ligaments:

The **right triangular ligament** extends from the right crus of the diaphragm above the central tendinous part to the right lateral lobe of the liver.

The **left triangular ligament** extends from the left crus of the diaphragm to the left lateral lobe of the liver.

The **coronary ligament** is the peritoneal reflection which attaches the liver to the diaphragm. On the right it is continuous with the right triangular ligament and on the left it is continuous with the left triangular ligament. Ventrally, right and left parts of the coronary ligament converge to form the falciform ligament.

The **falciform ligament** extends from the liver to the diaphragm and ventral abdominal wall to the umbilicus. The round ligament of the liver, which is the remnant of the **umbilical vein** of the fetus, may be found in the young animal as a small fibrous cord lying in the free edge of the falciform ligament. It enters the fissure for the round ligament of the liver. In adult dogs the falciform ligament is filled with fat and it persists only from the diaphragm to the umbilicus.

VESSELS AND NERVES OF THE ABDOMINAL VISCERA

VAGAL NERVES

The paired **vagus nerve** (Fig. 129) carries preganglionic parasympathetic efferent fibers to all of the thoracic and most of the abdominal organs. Right and left vagus nerves divide behind the roots of the lungs into **dorsal and ventral branches.** The dorsal branches unite near the diaphragm to form the dorsal vagal trunk. The ventral branches unite behind the roots of the lungs to form the ventral vagal trunk. These trunks lie on the dorsal and ventral surfaces of the terminal part of the esophagus. From the trunks, as well as from the dorsal and ventral branches of the vagi, nerves arise to supply the esophagus. The vagal trunks pass through the diaphragm and course along the lesser curvature of the stomach.

Transect the left crus of the diaphragm at the esophageal hiatus and reflect it to expose the vagal trunks. Observe the **ventral vagal trunk.** It supplies the liver, the parietal surface of the stomach and the pylorus.

The **dorsal vagal trunk** (Fig. 129) gives off a **celiac branch** which passes dorsocaudally and contributes to the formation of the celiac and cranial mesenteric plexuses. The branches in these plexuses follow the terminal branching of the respective blood vessels to the intestines as far caudally as the left colic flexure. The dorsal vagal trunk continues along the lesser curvature of the stomach to supply the visceral surface and the pylorus.

SYMPATHETIC TRUNK

To expose the caudal thoracic and lumbar parts of the sympathetic trunk on the left side reflect the abdominal and caudal thoracic walls so that the kidney and the crura of the diaphragm are freely accessible. Reflect the kidney toward the median plane. The deep circumflex iliac artery and vein arise from the aorta in the caudal part of the lumbar region and may be transected and turned aside. These vessels will be described later with the branches of the abdominal aorta.

The psoas minor arises from the fascia of the muscle dorsal to it, the quadratus lumborum, and from the last thoracic and the first five lumbar vertebral bodies. It is ventral and medial to the quadratus lumborum and

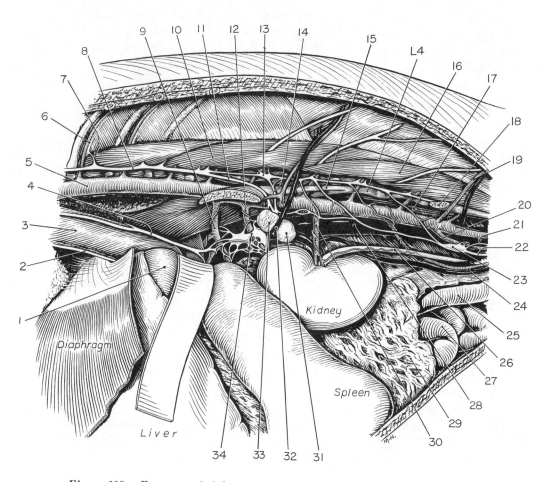

Figure 129. Exposure of abdominal autonomic nervous system on left side.

1. Stomach
2. Ventral trunk of vagus n.
3. Esophagus
4. Dorsal trunk of vagus n.
5. Aorta
6. Intercostal a. and n.
7. Ramus communicans
8. Sympathetic trunk
9. Celiac a.
10. Quadratus lumborum
11. Major splanchnic n.
12. Cranial mesenteric a.
13. Lumbar sympathetic ganglion at L2
14. Minor splanchnic n.
15. Left crus of diaphragm
16. Psoas major
17. Lumbar splanchnic n.
18. Transected psoas minor
19. Deep circumflex iliac a.
20. Caudal mesenteric a.
21. Left hypogastric n.
22. Caudal mesenteric plexus
23. Caudal mesenteric ganglion
24. Testicular a. and v.
25. Descending colon
26. Cranial ureteral a.
27. Jejunum
28. Caudal vena cava
29. Greater omentum
30. Renal a. and plexus
31. Adrenal gland
32. Phrenicoabdominal v.
33. Adrenal plexus
34. Celiac and cranial mesenteric ganglia and plexus.

psoas major (Fig. 129). It inserts on the ilium on the tubercle of the psoas minor dorsal to the iliopubic eminence. Transect the tendon of the psoas minor caudal to the deep circumflex iliac vessels. Remove the muscle from its vertebral origins.

SPLANCHNIC NERVES

The **splanchnic nerves** are sympathetic fibers which run between the sympathetic trunk and the abdominal autonomic ganglia.

The **major splanchnic nerve** (Fig. 129) leaves the sympathetic trunk at the level of the twelfth or thirteenth thoracic sympathetic ganglion. It passes dorsal to the crus of the diaphragm, enters the abdominal cavity and courses to the adrenal and celiacomesenteric plexuses.

The **minor splanchnic nerves** (Fig. 129), generally two, usually leave the last thoracic and first lumbar sympathetic ganglia. They supply nerves to the adrenal gland, ganglion and plexus and they terminate in the celiacomesenteric ganglia and plexus. Dissect the origin of these nerves and their course to the adrenal gland.

The **lumbar splanchnic nerves** (Fig. 129) arise from the second to the fifth lumbar sympathetic ganglia. In general they are distributed to the aorticorenal, cranial mesenteric, and caudal mesenteric plexuses. Observe the origin of these nerves.

ABDOMINAL NERVE PLEXUSES AND GANGLIA

In the abdomen, branches of the vagus and splanchnic nerves intermingle around the major abdominal arteries to form nerve plexuses. These plexuses are named according to the vessel upon which they are found. The plexuses supply the musculature of the artery and the viscera supplied by the branches of that artery.

Several sympathetic ganglia are located in the abdomen in close association with the plexuses. These ganglia are collections of cell bodies of postganglionic axons. Preganglionic axons in the splanchnic nerves must synapse in one of these ganglia. Preganglionic vagal axons (parasympathetic) do not synapse here but pass through the plexuses to the wall of the organ innervated, where they synapse on a cell body of a postganglionic axon.

The **celiac ganglia** (Fig. 129) lie on the right and left surfaces of the celiac artery close to its origin. They are often interconnected and numerous nerves from the ganglia follow the terminal branches of the celiac artery as a plexus.

The **cranial mesenteric ganglion** (Fig. 129) is unpaired and is located caudal to the celiac ganglion on the caudal surface of the cranial mesenteric artery, which it partly encircles. Most of its nerves continue distally on the cranial mesenteric artery as the cranial mesenteric plexus. Because of the close relationship of the celiac and cranial mesenteric plexuses and ganglia, they are referred to as the **celiacomesenteric ganglion and plexus.**

The **caudal mesenteric ganglion** (Fig. 129) is located ventral to the aorta around the caudal mesenteric artery, which is an unpaired branch of the aorta caudal to the kidneys. Lumbar splanchnic nerves enter the ganglion on each side. Branches may also come from the aortic and celiacomesenteric plexuses. Some of the nerves leaving the ganglion continue along the artery as the **caudal mesenteric plexus.**

The **right** and **left hypogastric nerves** leave the ganglion and course caudally near the ureters. They run in the mesocolon, incline laterally and disappear at the pelvic inlet. Their connections with the pelvic plexuses will be followed later.

BILIARY PASSAGES

Much of the biliary duct system within the liver is microscopic. The bile, which is secreted by the liver cells, is collected into the canaliculi, which drain into interlobular ducts. The interlobular ducts of each lobe unite to form **hepatic ducts** (Fig. 130), which emerge from each lobe. The arrangement of the hepatic ducts is variable.

The **gall bladder** (Fig. 130) is located in a fossa between the quadrate and right medial lobes of the liver. The full gall bladder extends through the liver and contacts the diaphragm (which is often stained green in preserved specimens). The neck of the gall bladder is continued as the **cystic duct.**

The main duct formed by the union of the hepatic ducts and the cystic

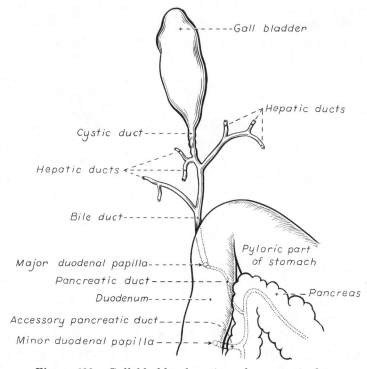

Figure 130. Gall bladder, hepatic and pancreatic ducts.

duct from the gall bladder is the **bile duct (ductus choledochus).** It terminates on the major duodenal papilla alongside the pancreatic duct. There are no valves in the biliary ducts and bile may flow in either direction. Observe the duct system in your specimen.

BRANCHES OF THE ABDOMINAL AORTA

1. The paired **lumbar arteries** (Figs. 131–133) leave the dorsal surface of the aorta. Each extends dorsally and terminates in a spinal and a dorsal branch. The spinal branches anastomose with the ventral spinal artery, which is within the vertebral canal and which supplies part of the spinal cord. The dorsal branches supply the muscles and skin above the lumbar vertebrae.

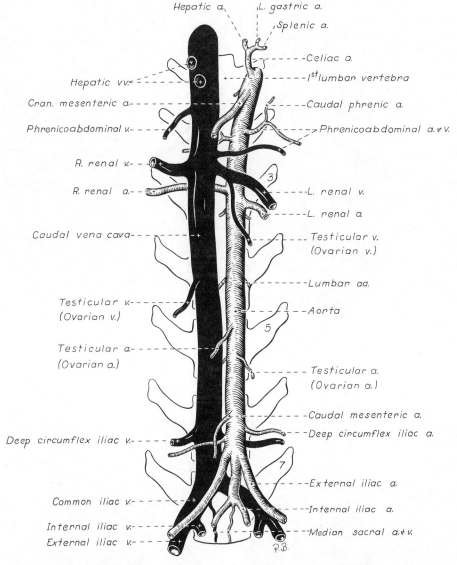

Figure 131. *Branches of abdominal aorta and tributaries of caudal vena cava.*

2. The **celiac artery** (Figs. 131, 132) arises from the aorta between the crura of the diaphragm. It has three branches: the hepatic artery, the left gastric artery and the splenic artery. The celiac plexus of nerves covers the artery in its course through the mesentery.

The **hepatic artery** (Figs. 131, 132) is the first branch to leave the celiac. Three to five branches leave the hepatic artery and enter the liver. The **cystic artery** leaves the last hepatic branch and supplies the gall

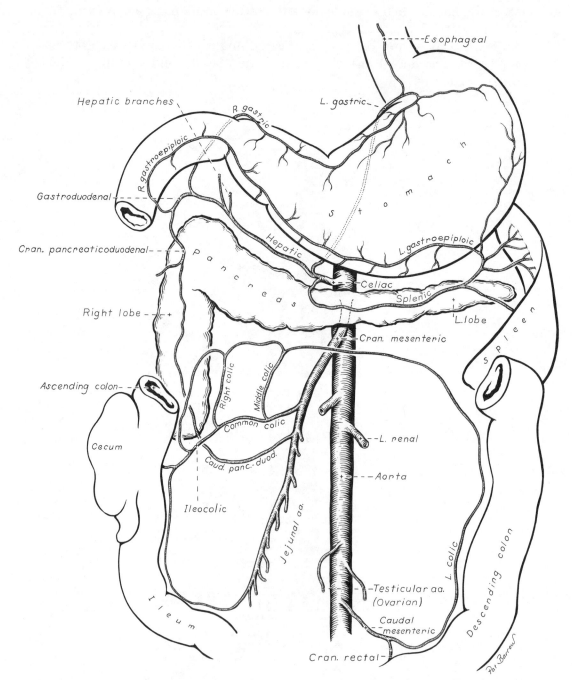

Figure 132. Visceral branches of aorta with principal anastomoses.

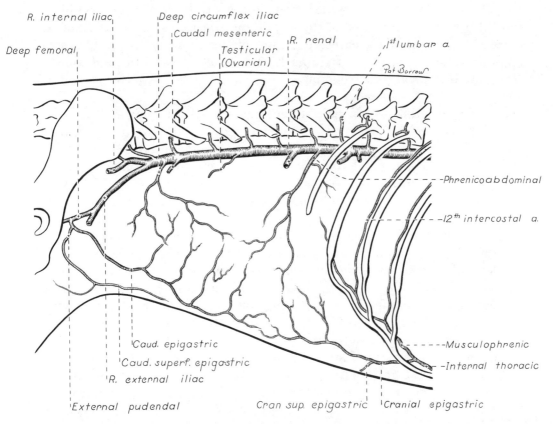

Figure 133. *Diagram of abdominal aorta.*

bladder. After giving off branches to the liver the hepatic artery terminates as the right gastric and the gastroduodenal arteries.

The **right gastric artery** is a small artery which extends from the pylorus toward the cardia to supply the lesser curvature of the stomach. It anastomoses with the left gastric artery. It need not be dissected.

The **gastroduodenal artery** supplies the pylorus and terminates as the right gastroepiploic and cranial pancreaticoduodenal arteries.

The **right gastroepiploic artery** enters and runs in the greater omentum along the greater curvature of the stomach. It supplies the stomach and the greater omentum. The right gastroepiploic artery anastomoses with the left gastroepiploic, a branch of the splenic artery.

The **cranial pancreaticoduodenal artery** follows the mesenteric border of the descending limb of the duodenum, where it supplies the duodenum and adjacent right lobe of the pancreas. It anastomoses with the caudal pancreaticoduodenal artery, which is a branch of the cranial mesenteric artery.

The **left gastric artery** (Figs. 131, 132) runs to the lesser curvature of the stomach, near the cardia, and supplies both surfaces of the stomach. It extends toward the pylorus, where it anastomoses with the right gastric artery. One or more **esophageal rami** run cranially on the esophagus.

The **splenic artery** approaches the hilus of the spleen near its middle, sends branches to each end of the organ and continues as the **left gastroepiploic** artery to the greater curvature of the stomach. The left gastroepiploic artery supplies the stomach and greater omentum and anastomoses with the right gastroepiploic.

3. The **cranial mesenteric artery** (Figs. 131, 132) leaves the aorta caudal to the celiac artery. It is surrounded by the cranial mesenteric plexus of nerves. Peripheral to the plexus are the mesenteric lymph nodes and branches of the portal vein. Reflect these from the vessel. Observe the branches of the cranial mesenteric artery.

The middle, right and ileocolic arteries arise from a common trunk from the cranial mesenteric artery.

The **middle colic artery,** the first branch from the common trunk, runs cranially in the mesocolon to the mesenteric border of the left colic flexure and descending part of the colon. It bifurcates near the left colic flexure. One branch runs distally in the descending mesocolon, supplies the descending colon and then anastomoses with the left colic artery, a branch of the caudal mesenteric artery. The other branch passes to the right and forms an arcade with the smaller right colic artery, and supplies the transverse colon.

The **right colic artery** runs in the right mesocolon toward the right colic flexure giving off branches to the distal part of the ascending colon and the adjacent transverse colon. It forms arcades with the middle colic artery and the colic branch of the ileocolic artery.

The **ileocolic artery** (Fig. 134) supplies the ileum, cecum and ascending colon. The ascending colon is supplied by the **colic branch.** The **cecal artery** crosses the dorsal surface of the ileocolic junction and supplies the cecum and ileum. The **ileal branch** passes to the mesenteric side of the ileum and supplies the cecum and ileum.

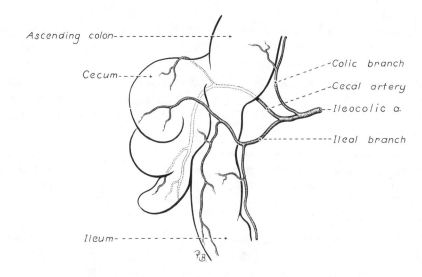

Figure 134. Ileocolic artery.

The **caudal pancreaticoduodenal artery** runs to the right in the mesoduodenum to the descending portion of the duodenum near the caudal flexure. It supplies the descending duodenum and the right lobe of the pancreas.

The **jejunal arteries** arise from the caudal side of the cranial mesenteric artery. They form arcades in the mesentery close to the jejunum.

The cranial mesenteric artery is terminated by **ileal arteries.**

4. The **phrenicoabdominal artery** is paired and arises from the aorta between the cranial mesenteric and renal arteries. The phrenicoabdominal artery crosses the ventral surface of the psoas muscles dorsal to the adrenal gland. The **caudal phrenic artery** runs cranially to supply the diaphragm. The **cranial abdominal artery** continues into the abdominal wall and ramifies between the transversus abdominis and the internal abdominal oblique.

The adrenal gland may receive branches from the aorta, caudal phrenic, renal or lumbar artery.

5. The **renal arteries** (Figs. 131, 132) leave the aorta at different levels. The right one arises cranial to the left, in conformity with the more cranial position of the right kidney. It is longer than the left and lies dorsal to the caudal vena cava.

6. The **ovarian artery** (Figs. 131, 132) of the female is homologous with the testicular artery of the male. This paired vessel arises from the aorta about halfway between the renal and external iliac arteries. The ovarian artery varies in size, position and tortuosity with the degree of development of the uterus. Each ovarian artery divides into two or more branches in the mesovarium just medial to the ovaries. Branches supply the ovary and its bursa and the uterine tube and horn. The branch to the uterine horn anastomoses with the uterine artery, which runs in the mesometrium.

The **testicular artery** leaves the aorta in the midlumbar region and crosses the ventral surface of the ureter. The testicular artery, vein and nerve plexus lie in a peritoneal fold, the **mesorchium,** which can be followed to the level of the vaginal ring.

7. The **caudal mesenteric artery** (Figs. 131, 132) is unpaired and arises near the termination of the aorta. It enters the descending mesocolon and runs caudoventrally to the mesenteric border of the descending colon, where it terminates in two branches of similar size. The **left colic artery** follows the mesenteric border of the descending colon cranially to anastomose with the middle colic artery. The **cranial rectal artery** descends along the rectum and anastomoses with the caudal rectal artery.

8. The **deep circumflex iliac artery** is paired and arises from the aorta close to the origin of the external iliac artery. It crosses the sublumbar muscles laterally, and at the lateral border of the psoas major it

supplies the musculature of the caudodorsal portion of the abdominal wall. The deep circumflex iliac artery perforates the abdominal wall and appears superficial ventral to the tuber coxae. It supplies the skin of the caudal abdominal area, the flank and the cranial thigh.

PORTAL VENOUS SYSTEM

The **portal vein** (Fig. 135) carries venous blood to the liver from the stomach, the small intestine, the cecum, the colon, the pancreas and the spleen. Separate the caudate process of the liver from the descending part of the duodenum and find the portal vein. Reflect the peritoneum and fat from its surface as far caudally as the root of the mesentery and expose its branches.

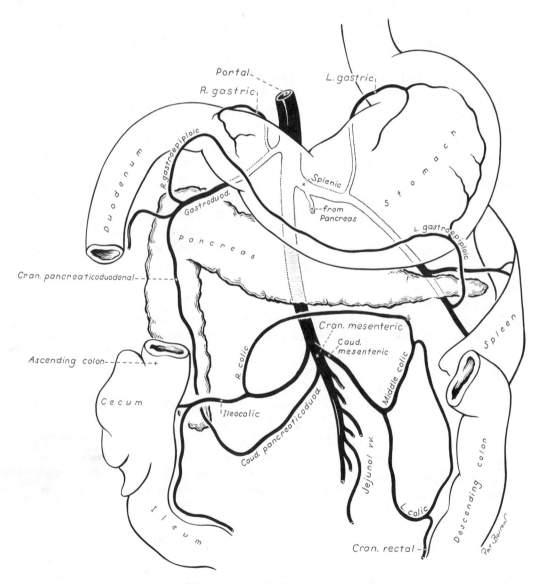

Figure 135a.   Tributaries of portal vein.

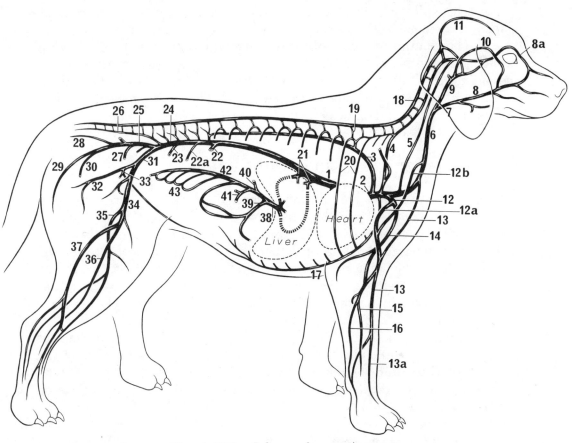

Figure 135b. *Schema of venous system.*

1. Caudal vena cava
2. Cranial vena cava
3. Azygos
4. Vertebral
5. Internal jugular
6. External jugular
7. Linguofacial
8. Facial
8a. Angularis oculi
9. Maxillary
10. Superficial temporal
11. Dorsal sagittal sinus
12. Axillary
12a. Axillobrachial
12b. Omobrachial
13. Cephalic
13a. Accessory cephalic
14. Brachial
15. Radial
16. Ulnar
17. Internal thoracic
18. Right vertebral venous plexus
19. Intervertebral
20. Intercostal
21. Hepatic

22. Renal
22a. Testicular or Ovarian
23. Deep circumflex iliac
24. Common iliac
25. Right internal iliac
26. Median sacral
27. Urogenital
28. Lateral caudal
29. Caudal gluteal
30. Internal pudendal
31. Right external iliac
32. Deep femoral
33. Pudendoepigastric trunk
34. Femoral
35. Medial saphenous
36. Cranial tibial
37. Lateral saphenous
38. Portal
39. Gastroduodenal
40. Splenic
41. Caudal mesenteric
42. Cranial mesenteric
43. Jejunal

1. The **gastroduodenal vein** is a small, proximal branch of the portal vein. It drains the pancreas, the stomach, the duodenum and the greater omentum.

2. The **splenic vein** is a large branch that receives blood from the spleen, the stomach, the pancreas and the greater omentum.

3. The **cranial** and **caudal mesenteric veins** are the distal terminal branches of the portal vein. The cranial mesenteric vein arborizes in the mesentery and collects blood from the jejunum, the ileum, the caudal duodenum and the right lobe of the pancreas. The caudal mesenteric vein drains the cecum and the colon.

PELVIC VISCERA, VESSELS AND NERVES

To remove the left pelvic limb reflect the penis and scrotum to the right. Cut through the pelvic symphysis with a cartilage knife or a saw. Locate the wing of the ilium and sever all muscles that attach to its medial and ventral surfaces. Apply even but constant lateral pressure to the left os coxae and at the same time cut through the sacroiliac joint. Cut the attachment of the penis to the left ischiatic tuberosity. Leave the limb attached. All structures more easily traced from the left should be dissected from this side.

The **levator ani** (Figs. 136, 143) muscle lies medial to the coccygeus muscle. It is a broad, thin muscle originating on the medial edge of the shaft of the ilium and the dorsal surface of the pubis and the pelvic symphysis. The muscle appears caudal to the coccygeus, where it inserts on caudal vertebrae 3 to 7. Transect this muscle on the left side near its origin and reflect it.

The **coccygeus muscle** (Fig. 136) lies lateral to the levator ani muscle. It arises from the ischiatic spine and inserts on the transverse processes of caudal vertebrae 2 to 4. Transect the muscle at its origin and remove the left pelvic limb. The levator ani and coccygeus muscles of each side form a pelvic diaphragm through which the genitourinary and digestive tracts open to the outside.

The **pelvic plexus** (Fig. 136) lies caudal to a plane passing through the pelvic inlet and dorsal to the prostate. It is closely applied to the surface of the rectum and can be identified by tracing the left hypogastric nerve to it. Occasionally ganglia are large enough to be recognized in the plexus. The pelvic plexus contains sympathetic fibers from the hypogastric nerve and parasympathetic fibers from the pelvic nerve.

The **pelvic nerve** is formed by parasympathetic preganglionic axons that leave the sacral nerves. Find the left pelvic nerve on the lateral wall of the distal portion of the rectum. Trace it proximally to its origin. It supplies branches to the urogenital organs, the rectum and the descending colon.

The pelvic part of the peritoneal cavity extends caudally on all sides of the rectum. On the ventral surface of the rectum the peritoneum is reflected on the vagina or prostate. The mesorectum divides the dorsal extension of this cavity into bilateral parts. Its most caudal extension is generally to a transverse plane through the second caudal vertebra.

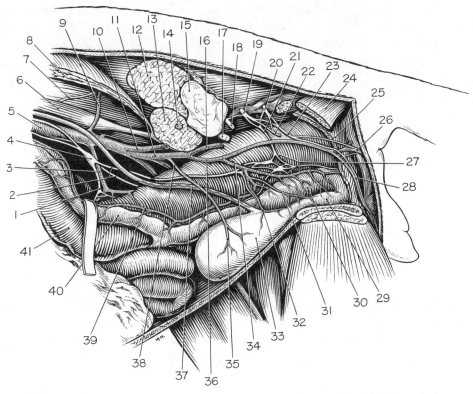

Figure 136. *Autonomic nerves and vessels of pelvic region, left lateral view.*

1. Caudal mesenteric plexus
2. Right and left hypogastric nerves
3. Caudal mesenteric artery
4. Caudal mesenteric ganglion
5. Aorta
6. Psoas minor
7. Lateral cutaneous femoral nerve
8. Abdominal oblique muscles
9. Deep circumflex iliac artery
10. External iliac artery
11. Internal iliac artery
12. Quadratus lumborum
13. Iliopsoas
14. Femoral nerve
15. Sacroiliac articulation
16. Caudal gluteal artery
17. Lumbar nerves 6 and 7
18. First sacral nerve
19. Second sacral nerve
20. Third sacral nerve
21. Pelvic nerve
22. Caudal cutaneous femoral nerve
23. Pudendal nerve
24. Coccygeus
25. Levator ani
26. Perineal nerve and artery
27. Pelvic plexus
28. Artery and nerve to clitoris
29. Urethra
30. Vagina
31. Urethral branch of urogenital artery
32. Caudal vesical artery
33. Bladder
34. Urogenital artery
35. Cranial vesical artery
36. Internal pudendal artery
37. Ureter and ureteral branch of urogenital artery
38. Umbilical artery
39. Uterine artery
40. Uterine horn
41. Descending colon

ILIAC ARTERIES

The paired **iliac arteries** (Figs. 131, 133, 136) supply the pelvis and hind leg. The **external iliac** runs ventrocaudally and becomes the femoral artery which leaves the abdomen through the vascular lacuna. The **internal iliac** arises caudal to the external iliac and passes caudolaterally into the pelvis.

The internal iliac artery and the smaller, unpaired **median sacral artery** terminate the aorta. Find the origin of these vessels. The internal iliac artery gives off the rudimentary umbilical artery and terminates cranial to the sacroiliac joint as the caudal gluteal and internal pudendal arteries. Dissect the following vessels on the left side.

In the fetus, the **umbilical artery** is a large, paired vessel that carries blood from the aorta to the placenta through the umbilicus. Find the remnant of this vessel which arises near the origin of the internal iliac artery and courses to the apex of the bladder in its lateral ligament. In some specimens it remains patent this far and supplies the bladder with cranial vesical arteries. Distal to the bladder the vessel is obliterated.

Find the origin of the **internal pudendal artery** (Figs. 136–138) from the internal iliac and dissect its branches. It is the smaller, more ventral branch which runs caudally on the terminal tendon of the psoas minor. At the level of the sacroiliac joint the internal pudendal gives rise to the urogenital artery.

The **urogenital artery** forms an angle of about 45 degrees with the internal pudendal. It passes ventrally in an arch and terminates as cranial and caudal branches. The **cranial branch** supplies a ureteral branch, a urethral branch, a caudal vesical artery and an artery to the

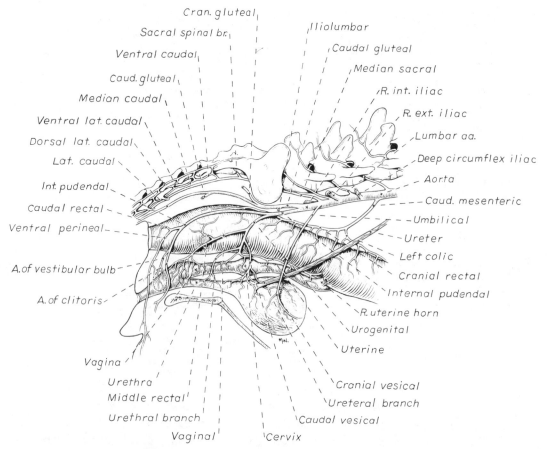

Figure 137.   Arteries of female pelvis, right lateral view.

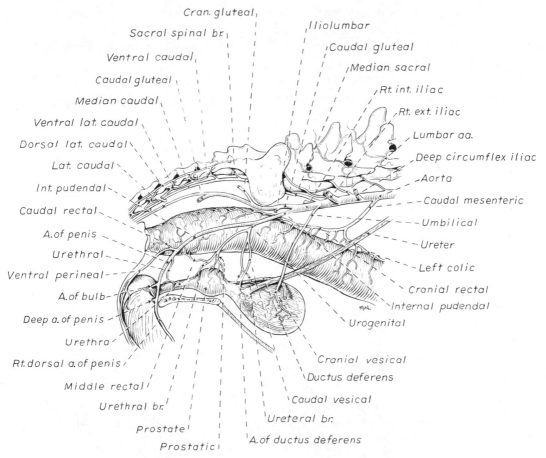

Figure 138. *Arteries of male pelvis, right lateral view.*

ductus deferens in the male or uterine artery in the female. The **caudal branch** supplies a urethral branch, a middle rectal artery and a prostatic artery in the male or vaginal artery in the female.

Remove the skin and fat from the right ischiorectal fossa. The internal pudendal artery (Fig. 136) continues along the dorsal border of the ischiatic spine lateral to the coccygeus muscle and medial to the gluteal muscles and reaches the ischiorectal fossa. Here it terminates as a ventral perineal artery and an artery of the penis or clitoris. These vessels may be dissected on either side.

The **ventral perineal artery** may be seen passing caudoventrally. It supplies a branch to the rectum and anus and terminates in the skin of the perineum and the scrotum or vulva.

The **artery of the penis** courses caudoventrally and terminates as three branches: The **artery of the bulb of the penis** arborizes in the bulb, and continues to supply the corpus spongiosum and penile urethra. Observe this artery as it enters the bulb. The **deep artery of the penis** enters the corpus cavernosum at the crus. The **dorsal artery of the penis** runs on the dorsal surface to the level of the bulbus glandis where it divides and sends branches to the prepuce and pars longa glandis. The penile arteries are accompanied by veins which have an important role in the mechanism of erection (Fig. 139).

In the female, the **artery of the clitoris** courses caudoventrally to supply the clitoris and vestibular bulb.

The **urinary bladder** (Figs. 140–142) has an apex, a body and a neck. Three peritoneal folds, ligaments, are reflected from the bladder upon the

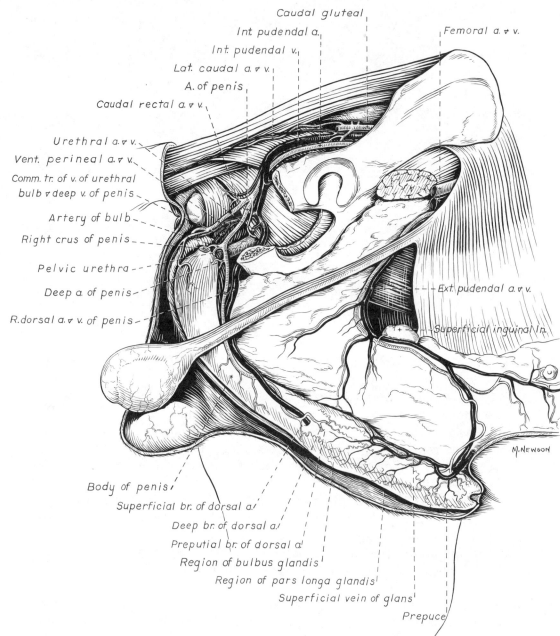

Figure 139.   *Vessels of penis and prepuce. (From Christensen, American Journal of Anatomy, 1954)*

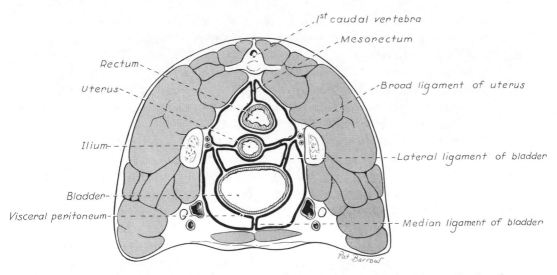

Figure 140. *Schematic cross-section of peritoneum at level of bladder in the female.*

pelvic and abdominal walls. The **median ligament of the bladder** leaves the ventral surface of the bladder and attaches to the abdominal wall as far cranial as the umbilicus. In the fetus it contains the urachus and umbilical arteries. The **lateral ligament of the bladder** passes to the pelvic wall.

Observe the pattern of the bundles of smooth muscle on the surface of the bladder. They pass obliquely across the neck of the bladder and the origin of the urethra. No sphincter is present in the bladder.

The **urethral muscle** is striated and is confined to the pelvis, where it surrounds the pelvic urethra and serves as a sphincter. Make a midventral incision through the bladder wall and the urethra.

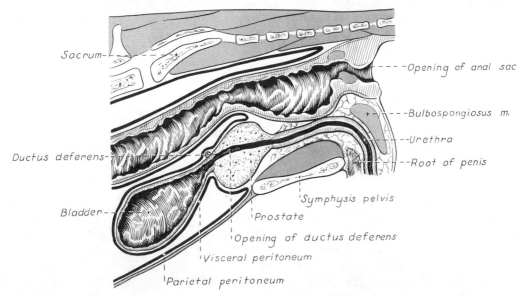

Figure 141. *Schematic median section of peritoneum through male pelvic region.*

Examine the mucosae of the bladder and urethra. If the bladder is contracted its mucosa will be thrown into numerous folds, or **rugae,** as a result of its inelasticity.

Observe the entrance of the ureters into the bladder. These lie opposite each other near the neck of the organ. The **trigone of the bladder** is the dorsal triangular area located within lines connecting the ureteral openings into the bladder and the urethral exit from it.

The **rectum** (Fig. 141) continues the descending colon through the pelvis. It begins at the pelvic inlet. The **anal canal** is a continuation of the rectum to the anus. The anal canal begins where the mucosa of the rectum forms longitudinal folds. Distal to these folds is the cutaneous zone of the anal canal which has fine hairs, glands and, on each side, the prominent openings of the anal sacs. The anal canal is surrounded by both a smooth **internal** and a striated **external sphincter muscle.** The external opening of the anal canal is the **anus.** Transect the external sphincter on the left and reflect it from the anal sac. This muscle receives its nerve supply from the caudal rectal nerve and its blood supply from the perineal arteries.

Observe the **anal sac,** expose its duct and find the opening in the cutaneous zone of the anal canal. In the wall of this sac are microscopic glands whose secretion accumulates in the lumen of the sac and is discharged through the duct of the anal sac. Open the sac and examine its interior.

The **internal sphincter muscle** of the anus is an enlargement of the smooth circular muscle coat of the anal canal. It is not as large as the external sphincter, under which it lies. Reflect the anal sac and observe the internal sphincter.

The **rectococcygeus muscle** continues the longitudinal coat of the rectum to the ventral surface of the tail. Reflect the levator ani and coccygeus muscles from the left side of the rectum. Observe the rectococcygeus muscle arising from the dorsal surface of the rectum cranial to the sphincter muscles. Trace it caudally to its insertion on the caudal vertebrae.

MALE:

The **prostate gland** (Figs. 141, 142) completely surrounds the neck of the bladder and the beginning of the urethra. Examine the surface, form, size and location of the prostate on several specimens. The normal size and weight of the prostate varies greatly. The organ generally lies at the brim of the pubis. The prostate is flattened dorsally and rounded ventrally and on the sides. It is heavily encapsulated. Muscle fibers from the bladder run caudally on its dorsal surface. A longitudinal septum leaves the ventral part of the capsule and reaches the urethra, thus partially dividing the gland ventrally into right and left lobes. This is indicated on the ventral surface by a shallow but distinct longitudinal furrow. Notice that the urethra runs through the center of the gland. Continue the

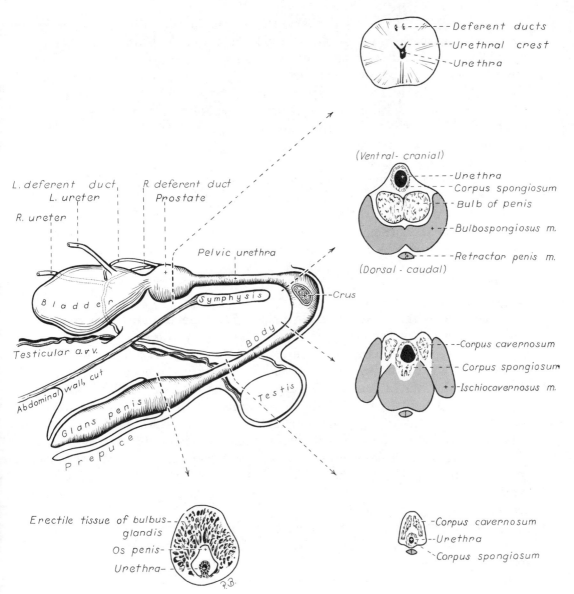

Figure 142. Schema of male urogenital system.

incision that opened the bladder through the prostate and urethra to expose the dorsal wall of the urethra.

The male **urethra** is composed of a **pelvic part** within the pelvis and a **spongy part** within the penis. The **urethral crest** protrudes into the lumen from the dorsal wall of the prostatic urethra. Near its middle and protruding farthest into the lumen of the urethra is a hillock, the **colliculus seminalis.** On each side of this eminence the deferent ducts open. Many prostatic openings are found on both sides of the urethral crest and can usually be seen if the gland is compressed.

The **penis** is composed of a root, a body and a glans (Fig. 142). The dorsal surface of the penis faces the pelvic symphysis and the abdominal wall. In the non-erect state the glans is entirely withdrawn into the prepuce.

The **prepuce** is a tubular sheath or fold of integument which is continuous with the skin of the ventral abdominal wall and is reflected over the glans. It has a smooth internal layer and a haired external layer which meet at the **preputial orifice.** At its deepest recess, the **fornix of the prepuce,** the internal layer is reflected onto the glans. In the erect state the fornix is eliminated since the internal layer of the prepuce is closely applied to the body of the penis. Open the prepuce by a midventral incision from the orifice to the fornix. Continue the midventral incision to the anus so as to expose the entire length of the penis.

The **root** of the penis is formed by right and left crura which originate on the ischiatic tuberosity of each side. The root ends where the crura blend with each other on the midline. Each **crus** is composed of cavernous tissue (corpus cavernosum penis) surrounded by a thick fibrous tunic (tunica albuginea). Examine the root. The left crus was transected when the limb was removed and the trabeculae and vascular spaces can be seen on the cut surface. Note the firm attachment of the right crus to the ischiatic tuberosity.

The **ischiocavernosus muscle** arises from the ischiatic tuberosity, covers the origin of the crus and inserts distally on the crus.

The narrow **retractor penis muscles** (Fig. 143) originate from the ventral surface of the sacrum or first two caudal vertebrae and extend distally on the ventral surface of the penis to the level of the glans, where they insert. In the region of the anal sphincter there is muscle fiber exchange between the retractor penis muscle and the external anal sphincter. Observe the retractor muscle on the body of the penis.

The **bulbospongiosus muscle** bulges between the ischiocavernosus muscles ventral to the external anal sphincter. The fibers of the bulbospongiosus are transverse and cover the **bulb** of the penis and the proximal end of the body of the penis.

The **corpus cavernosum penis** of each crus converges toward its fellow on the dorsal side of the body of the penis, and the two corpora extend side by side throughout the body to the os penis. A median septum completely separates the two corpora, and each is covered by a white

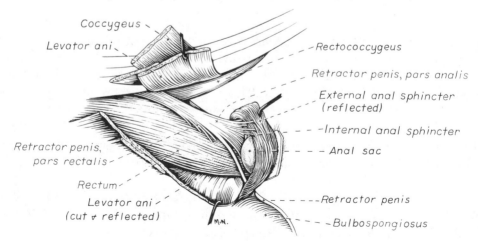

Figure 143.   *Muscles of the anal region, lateral aspect.*

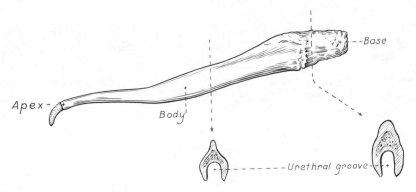

Figure 144. *Os penis, lateral view, with cross-sections.*

capsule (tunica albuginea) throughout its length. Make several incomplete transections through the body of the penis to study these structures.

The **os penis** (Figs. 142, 144) is a long, ventrally grooved bone which lies almost entirely within the glans penis. The expanded, rough, truncate base of the bone originates in the tunica albuginea at the distal end of the corpora cavernosa. The body of the os penis extends through the glans penis. The base and body are grooved ventrally by the **urethral groove,** which surrounds the urethra and the corpus spongiosum on three sides. The bone ends as a long, pointed fibrocartilage in the tip of the glans, dorsal to the urethral opening.

The **corpus spongiosum penis** consists of a median **bulbus penis** at the level of the ischial arch, and a layer of cavernous tissue around the urethra which extends from the bulb to the external urethral orifice. At the level of the collar-like bulbus glandis of the glans of the penis there is a communication between the corpus spongiosum penis and the **corpus spongiosum glandis.**

The **glans** of the penis is composed of two parts, the proximal **bulbus glandis** and the distal, more elongate **pars longa glandis.** The bulbus glandis, surrounding the proximal end of the os penis, is an expansile vascular structure that is largely responsible for retaining the penis within the vagina during copulation. The pars longa glandis is a cavernous tissue structure that overlaps the distal half of the bulbus glandis and continues to the distal end of the penis, partially encircling the os penis and the urethra. The pars longa glandis is separated from the bulbus glandis by a layer of connective tissue. Venous channels drain the pars longa glandis into the bulbus glandis through this layer. Make a longitudinal incision on the dorsum of the penis through the glans to observe its structure.

FEMALE:

The **cervix** (Fig. 145) is the constricted caudal portion of the uterus. The **cervical canal** is in a nearly vertical position, with the uterine opening (**internal uterine ostium**) dorsal, and the vaginal opening (**external uterine ostium**) ventral in position.

The **vagina** (Fig. 145) is located between the uterine cervix and the vestibule. The most cranial part of the vagina is the **fornix,** which extends cranial to the cervix along its ventral margin. The mucosal lining of the remaining part of the vagina is thrown into longitudinal folds which have small transverse folds. These are evidence of its ability to enlarge in both diameter and length. The longitudinal folds end dorsally at the level of the urethral orifice where the vagina joins the vestibule.

The **vestibule** (Fig. 145) is the cavity which extends from the vagina to the vulva. The **urethral crest** projects from the floor of the cranial part of the vestibule. The urethra opens on this tubercle.

In the floor of the vestibule, deep to the mucosa, are two elongate masses of erectile tissue, the **vestibular bulbs.** They are homologous to the bulb of the penis of the male and lie in close proximity to the body of the clitoris.

The **clitoris** is the female homologue of the penis. It is a small structure located in the floor of the vestibule near the vulva. It is composed of paired crura, a short body and a glans clitoridis. The glans clitoridis is a very small erectile structure that lies in the fossa clitoridis. The fossa is a depression in the floor of the vestibule and should not be mistaken as the urethral opening. The dorsal wall of the fossa partly covers the glans clitoridis and is homologous to the male prepuce. Identify the fossa and glans clitoridis. Only rarely is an os clitoridis present in the glans.

The **vulva** includes the two **labia** and the external urogenital orifice that they bound, the **rima pudendi.** The labia fuse above and below the

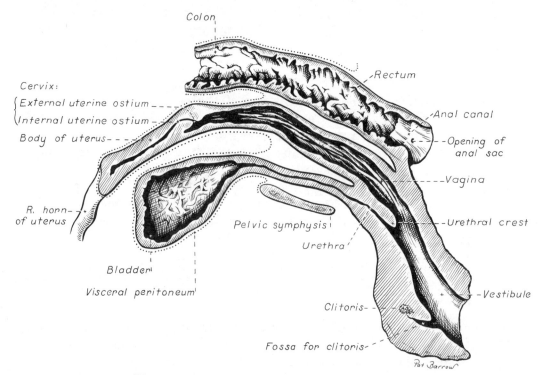

Figure 145. Female pelvic viscera, median section.

rima pudendi to form **dorsal** and **ventral commissures.** The dorsal commissure is at or slightly below a dorsal plane through the pelvic symphysis. The ventral commissure is directed caudoventrally.

Observe the course of the female urethra by passing a flexible probe through it. It extends from the bladder caudodorsally over the cranial edge of the pelvic symphysis to the genital tract caudal to the vaginovestibular junction. It ends at the external urethral orifice on the urethral crest.

VESSELS AND NERVES OF THE PELVIC LIMB

Internal Iliac Artery

Make a skin incision on the medial surface of the right thigh to the stifle. Encircle the stifle and reflect the skin from the thigh, rump and proximal half of the tail.

Expose the insertion of the superficial gluteal deep to the proximal edge of the biceps femoris. Transect the insertion at this level. Reflect the proximal portion of the superficial gluteal to its origin. Transect the middle gluteal muscle 1 cm. from the crest of the ilium. Start at the cranial border of the bone and detach the muscle from the gluteal surface.

The **caudal gluteal artery** (Fig. 148) is the larger of the two terminal branches of the internal iliac artery. It arises opposite the sacroiliac joint and passes caudally toward the ischiatic spine parallel to the internal pudendal artery. The branches of the caudal gluteal are the iliolumbar, cranial gluteal, lateral caudal and dorsal perineal arteries (Fig. 146). The veins (Fig. 147) will not be dissected.

The **iliolumbar artery** (Fig. 148) arises close to the origin of the caudal gluteal artery or directly from the internal iliac. It passes across the cranioventral border of the ilium and supplies the psoas minor, iliopsoas, sartorius, tensor fasciae latae and middle gluteal muscles. On the lateral side observe its distribution to the deep surface of the cranial end of the middle gluteal.

The **cranial gluteal artery** and **nerve** enter the middle gluteal muscle by passing across the cranial part of the greater ischiatic notch of the ilium.

Transect the biceps femoris midway between its origin and the stifle. Transect the semitendinosus 1 cm. distal to the transection through the biceps. Turn both muscles toward their origins. The caudal gluteal artery lies on the ventrocranial side of the sacrotuberous ligament and in this location gives off several small branches to adjacent muscles: the lateral caudal artery to the tail and the dorsal perineal artery to the perineum. These need not be dissected.

Follow the caudal gluteal artery as it continues toward the ischiatic tuberosity in relation to the sacrotuberous ligament and the ischiatic nerve. Here it supplies the superficial and middle gluteals, the rotators of the hip, and the adductor muscle. It passes over the ischiatic spine with the ischiatic nerve and divides into several branches which supply the biceps femoris, the semitendinosus and semimembranosus muscles. Turn

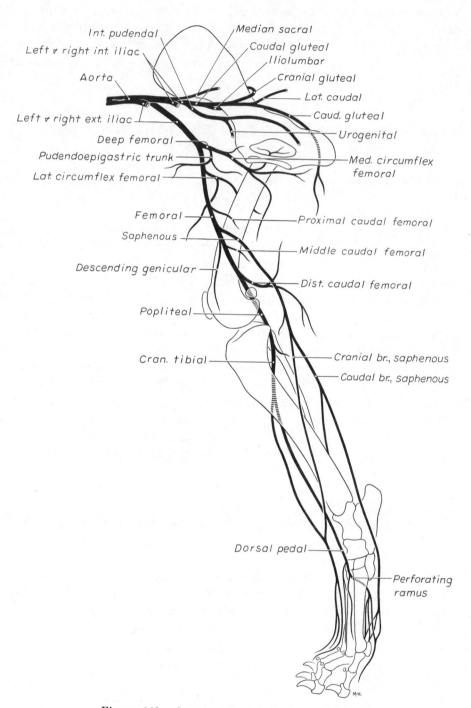

Figure 146. *Arteries of pelvic limb, medial view.*

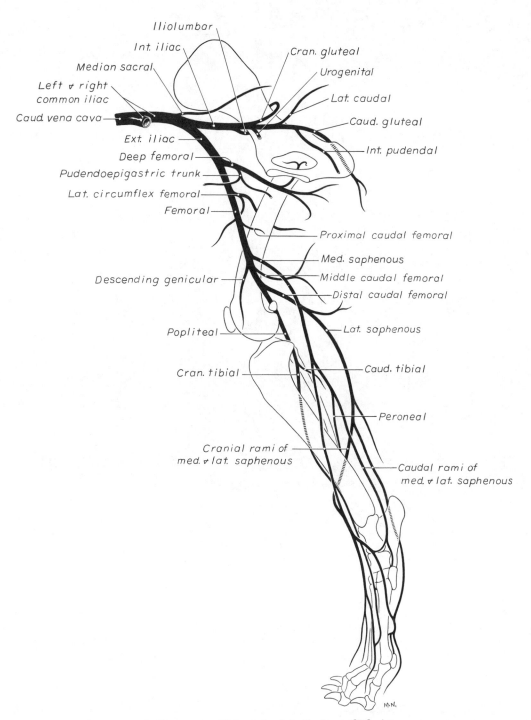

Figure 147. *Veins of pelvic limb, medial view.*

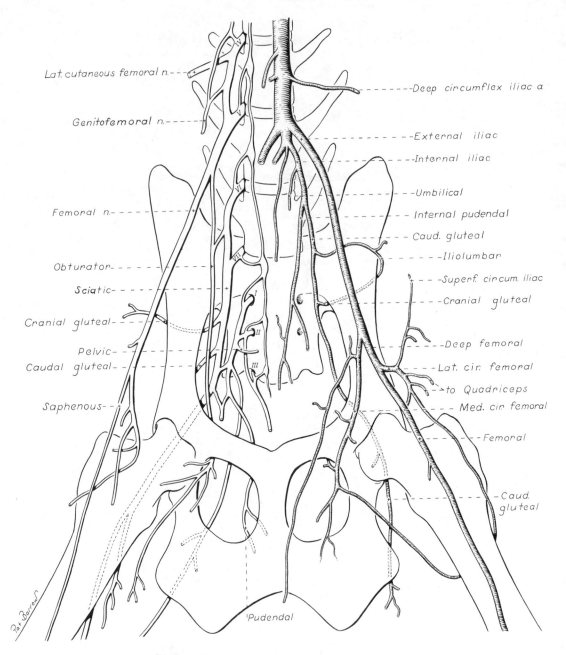

Figure 148. Lumbosacral nerves and arteries.

the biceps femoris caudally to expose the caudal gluteal artery lying deep to it close to the sacrotuberous ligament and ischiatic tuberosity.

EXTERNAL ILIAC ARTERY

The right **external iliac artery** (Fig. 148) arises from the aorta on a level with the sixth and seventh lumbar vertebrae. It runs caudoventrally and is related laterally near its origin to the common iliac vein and the psoas minor muscle. Farther distally it lies on the iliopsoas muscle. The external iliac upon passing through the abdominal wall becomes the femoral artery. The opening through which the external iliac artery passes is the vascular lacuna, located between the inguinal ligament and the pelvis.

The **deep femoral artery** is the only branch of the external iliac artery and arises inside the abdomen near the vascular lacuna. It passes caudally through the caudal part of the lacuna and enters the adductor muscles of the thigh. Two vessels leave the ventral surface of the deep femoral artery within the abdomen by a short **pudendoepigastric trunk.** These are the external pudendal and caudal epigastric arteries. The external pudendal artery has already been dissected.

The **caudal epigastric artery** (Fig. 149) arises from the pudendoepigastric trunk and passes cranially on the dorsal surface of the rectus ab-

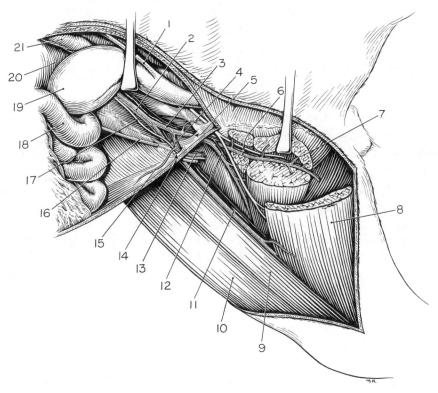

Figure 149.   *Deep femoral artery, medial view, pectineus removed and adductor transected.*

1. *Left ureter*
2. *External iliac artery*
3. *Deep femoral a. and v.*
4. *Pudendoepigastric trunk*
5. *Femoral lacuna*
6. *Transverse branch of medial circumflex femoral a. and v. and obturator n.*
7. *Adductor*
8. *Gracilis*
9. *Caudal part sartorius*
10. *Cranial part sartorius*
11. *Femoral a. and v.*
12. *Deep branch of medial circumflex femoral a. and v.*
13. *External pudendal a. and v.*
14. *Deep inguinal ring*
15. *Caudal epigastric artery*
16. *Round ligament of uterus*
17. *Genitofemoral nerve*
18. *Small intestine*
19. *Bladder*
20. *Rectum*
21. *Uterine horn*

dominis. It supplies the caudal half of the rectus abdominis and the ventral parts of the oblique and transverse muscles.

Expose the femoral artery and vein and the saphenous nerve in the **femoral triangle.** This triangle is bounded cranially by the sartorius, laterally by the vastus medialis and rectus femoris and caudally by the pectineus and adductor.

After giving off the pudendoepigastric trunk, the deep femoral artery is continued as the **medial circumflex femoral artery** (Fig. 149), which leaves the abdomen through the vascular lacuna and passes between the quadriceps femoris and pectineus muscles. Transect the pectineus at its origin and reflect it. Transect the origin of the gracilis and reflect the muscle caudally. Transect the origin of the adductor. Spare the branches of the medial circumflex femoral artery and obturator nerve that enter its cranial aspect. Remove portions of the adductor muscle to follow the distribution of the medial circumflex femoral artery.

As the medial circumflex femoral artery approaches the large adductor muscle it gives off a **deep branch** that descends distally between the adductor and vastus medialis muscles, both of which it supplies. Small branches supply the obturator muscles and the hip joint capsule. The **transverse branch** passes caudally through the adductor muscle, which it supplies, and terminates in the semimembranosus muscle.

The **femoral artery** is the continuation of the external iliac artery beyond the level of the vascular lacuna. The branches of the femoral artery in the order that they arise are superficial circumflex iliac, lateral circumflex femoral, proximal caudal femoral, saphenous, descending genicular and middle and distal caudal femoral.

The **superficial circumflex iliac artery** (Fig. 148) arises from the lateral side of the femoral artery near or with the lateral circumflex femoral artery. The superficial circumflex iliac artery supplies both parts of the sartorius, the tensor fascia lata and the rectus femoris, and becomes superficial at the cranial ventral iliac spine of the tuber coxae. Transect both parts of the sartorius over the vessel and observe its branches.

The **lateral circumflex femoral artery** (Fig. 148) passes between the rectus femoris and the vastus medialis. Although most of the vessel arborizes in the quadriceps, it also supplies the tensor fasciae latae, the superficial and middle gluteals and the hip joint capsule.

The **proximal caudal femoral artery** leaves the caudal surface of the femoral just distal to the origin of the lateral circumflex femoral. It extends distocaudally over the pectineus and adductor muscles, which it supplies, and enters the deep surface of the gracilis.

Make a skin incision from the stifle to the claw of the second digit. Remove the skin as far distally as the metatarsal pad.

The **saphenous artery** (Figs. 146, 150), vein and nerve cross the medial surface of the semimembranosus. They continue distally between the converging borders of the caudal part of the sartorius and gracilis. Observe that the saphenous artery arises from the femoral proximal to the stifle. The saphenous artery supplies the skin on the

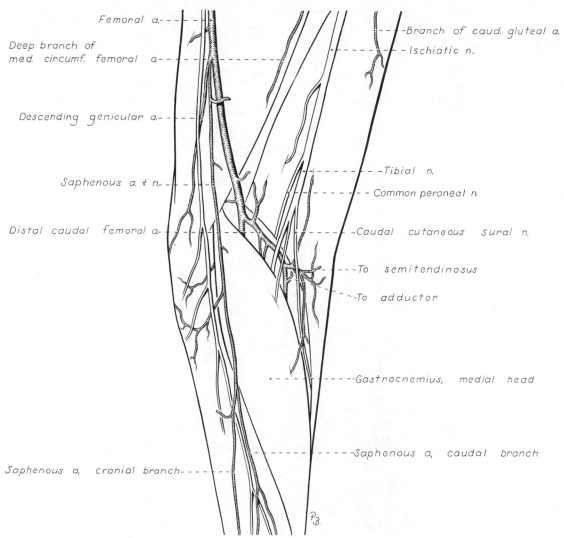

Femoral a.

Deep branch of
med. circumf. femoral a.

Descending genicular a.

Saphenous a. + n.

Distal caudal femoral a.

Branch of caud. gluteal a.

Ischiatic n.

Tibial n.

Common peroneal n.

Caudal cutaneous sural n.

To semitendinosus

To adductor

Gastrocnemius, medial head

Saphenous a, caudal branch

Saphenous a, cranial branch

Figure 150. *Arteries and nerves of right stifle region, medial view.*

medial side of the stifle and terminates in a cranial and a caudal branch.

The **cranial branch** (Figs. 146, 150) of the saphenous artery arises opposite the proximal end of the tibia, obliquely crosses the medial surface of the tibia and passes distally on the cranial tibial muscle. It crosses the flexor surface of the tarsus with this muscle. At the proximal part of the metatarsus it terminates as the dorsal common digital arteries.

The **caudal branch** of the saphenous artery arises at the proximal end of the tibia. It lies between the medial head of the gastrocnemius and the tibia. Distally it is related to the flexors of the digits and, with the tibial nerve, it crosses the medial plantar surface of the tarsus to enter the metatarsus. The vessel supplies branches to the tarsus and the deep structures of the proximal end of the metatarsus.

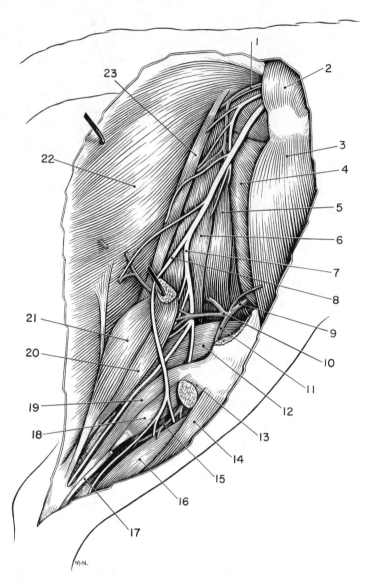

Figure 151.   *Arteries and nerves of right stifle region, lateral view.*

1. Caudal gluteal a.
2. Superficial gluteal
3. Vastus lateralis
4. Adductor
5. Semimembranosus
6. Semitendinosus
7. Tibial n.
8. Peroneal n.
9. Femoral a.
10. Distal caudal femoral a.
11. Popliteal a.
12. Gastrocnemius, medial head
13. Peroneus longus
14. Cranial tibial
15. Cranial tibial a.
16. Long digital extensor
17. Peroneus longus
18. Lateral digital extensor
19. Deep digital flexor, lateral head
20. Superficial digital flexor
21. Gastrocnemius, lateral head
22. Biceps femoris (reflected)
23. Caudal crural abductor

In the metatarsus it supplies branches that terminate as plantar common digital and plantar metatarsal arteries. These branches need not be dissected.

After the saphenous artery arises from the femoral, the latter disappears caudal or lateral to the semimembranosus. Turn the distal end of the semimembranosus craniomedially and trace the femoral artery to the gastrocnemius muscle.

The **descending genicular artery** (Figs. 146, 150) arises from the femoral distal to the origin of the saphenous and supplies the medial surface of the stifle.

The **middle caudal femoral artery** arises distal to the descending genicular and saphenous arteries and ramifies in the adductor and semimembranosus muscles.

The **distal caudal femoral artery** (Figs. 150-152) arises from the caudal surface of the last centimeter of the femoral. The femoral is continued by the popliteal artery on entering the gastrocnemius. Expose the distal caudal femoral artery and its branches, which supply the biceps femoris, the semimembranosus, the semitendinosus, the gastrocnemius and the digital flexors.

Popliteal Artery

Reflect the insertions of the gracilis, the semimembranosus and the semitendinosus to uncover the medial head of the gastrocnemius muscle. Transect the medial head of the gastrocnemius and reflect it.

The **popliteal artery** (Figs. 151, 152), a continuation of the femoral, passes between the two heads of the gastrocnemius muscle, crosses the medial surface of the superficial digital flexor muscle and courses over the flexor surface of the stifle. It inclines laterally under the popliteus muscle and perforates the lateral head of the deep digital flexor to reach the interosseous space. The popliteal artery supplies the stifle, gastroctibial arteries. The caudal tibial artery is a small vessel which leaves the caudal surface of the popliteal in the interosseous space.

Transect the popliteus where it covers the popliteal artery and follow the artery to the interosseous space.

The **cranial tibial artery** (Figs. 151, 152) passes between the tibia and the fibula. Reflect the fascia on the cranial aspect of the stifle and leg where it serves for the insertion of the biceps femoris. Separate the cranial tibial and long digital extensor muscles throughout their length. Observe the cranial tibial artery between these two muscles. Transect the peroneus longus muscle at its origin and reflect it to expose the cranial tibial artery emerging from the interosseous space between the tibia and fibula. It supplies the long digital extensor and cranial tibial muscles. The termination of the artery will be dissected with the peroneal nerves.

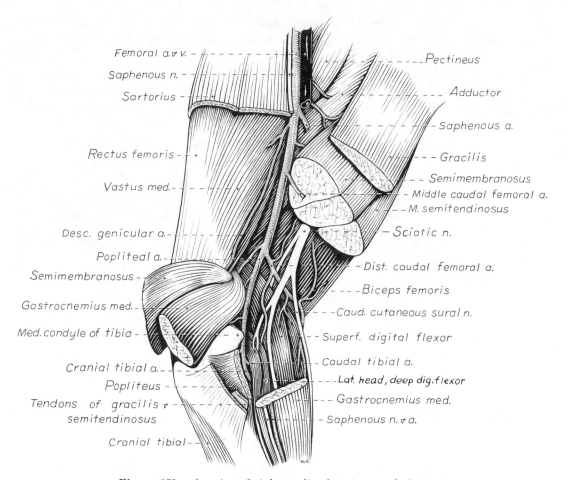

Figure 152. *Arteries of right popliteal region, medial view.*

LUMBOSACRAL PLEXUS

The **lumbosacral plexus** consists of the ventral branches of the lumbar and sacral nerves. Of the nerves that arise from this plexus, the cranial and caudal iliohypogastric, the ilioinguinal, the lateral cutaneous femoral and the genitofemoral have been dissected. The remainder will be dissected in the order of their accessibility.

1. The **obturator nerve** (Figs. 153, 154) arises from the fourth, fifth and sixth lumbar nerves. It is formed within the caudomedial portion of the iliopsoas muscle. It leaves the muscle dorsomedially, runs caudoventrally along the shaft of the ilium, penetrates the medial side of the levator ani muscle and leaves the pelvis by passing through the cranial part of the obturator foramen. It supplies the adductor muscles of the limb: the external obturator, the pectineus, the gracilis and the adductor. Locate this nerve on the medial side of the right ilium. Find it as it emerges ventrally from the obturator foramen and arborizes in the adductor muscles.

2. The **femoral nerve** (Figs. 153, 154) arises primarily from the

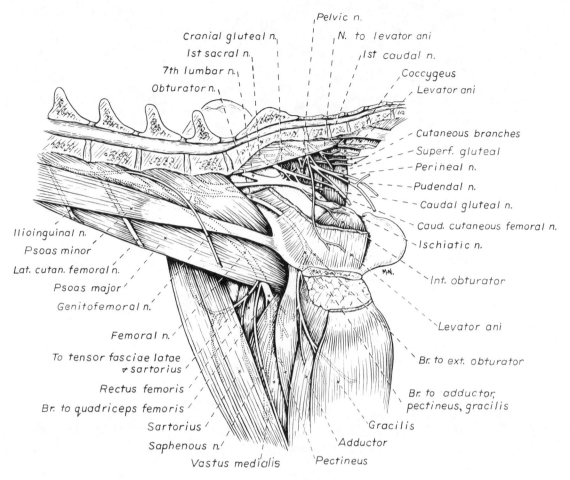

Cranial gluteal n.
1st sacral n.
7th lumbar n.
Obturator n.

Pelvic n.
N. to levator ani
1st caudal n.
Coccygeus
Levator ani

Cutaneous branches
Superf. gluteal
Perineal n.
Pudendal n.
Caudal gluteal n.
Caud. cutaneous femoral n.
Ischiatic n.

Ilioinguinal n.
Psoas minor
Lat. cutan. femoral n.
Psoas major
Genitofemoral n.

Int. obturator

Femoral n.
To tensor fasciae latae
v sartorius
Rectus femoris
Br. to quadriceps femoris
Sartorius
Saphenous n.
Vastus medialis

Levator ani

Br. to ext. obturator

Br. to adductor,
pectineus, gracilis

Gracilis
Adductor
Pectineus

Figure 153. Lumbosacral plexus, medial view.

fourth, fifth and sixth lumbar nerves. Find the femoral nerve with the lateral circumflex femoral artery. Observe its emergence from the iliopsoas muscle. Within the iliopsoas, the **saphenous nerve** arises from the cranial side of the femoral nerve. The saphenous or femoral innervates the sartorius muscle. The cutaneous portion of the saphenous supplies the skin on the medial side of the thigh, the stifle, leg, tarsus and paw. Follow this nerve as far distally as the tarsus.

The femoral nerve supplies branches to the iliopsoas muscle. It enters the quadriceps muscle between the rectus femoris and vastus medialis and supplies all four heads of the quadriceps.

In the right ischiorectal fossa identify the coccygeus muscle. Transect and reflect the sacral attachment of the sacrotuberous ligament, the superficial gluteal muscle and the middle gluteal muscle, which cover the greater ischiatic notch. This exposes the caudal gluteal artery and the sciatic nerve. Deep to these are the internal pudendal artery and pudendal nerve and the ventral branches of the sacral nerves. These ventral branches emerge from the two pelvic sacral foramina and the sacrocaudal intervertebral foramen.

3. The **pudendal nerve** (Fig. 153) arises from all three sacral nerves. It passes to the pelvic outlet, where it lies lateral to the levator ani and

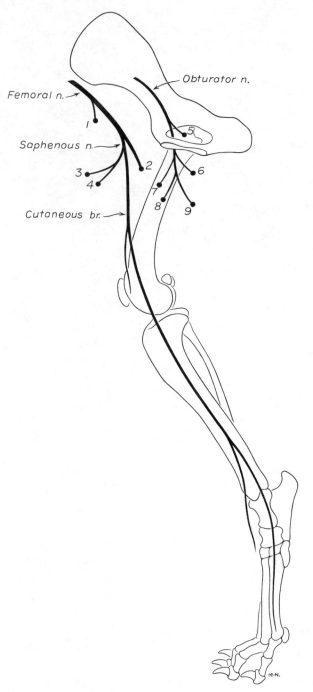

Figure 154. *Distribution of saphenous, femoral and obturator nerves.
Muscles innervated by numbered nerves.*
Femoral nerve
  1. Iliopsoas
  2. Quadriceps femoris
Saphenous nerve
  3. Sartorius, cranial part
  4. Sartorius, caudal part
Obturator nerve

  5. External obturator
  6. Adductor longus
  7. Pectineus
  8. Adductor magnus and brevis
  9. Gracilis

coccygeus muscles, medial to the superficial gluteal muscle and dorsal to the internal pudendal vessels. It appears superficially in the ischiorectal fossa after emerging from the medial side of the superficial gluteal muscle. The following branches arise from the pudendal nerve and should be located.

The **perineal nerves** arise from the dorsal surface of the pudendal nerve. They supply the skin of the anus, the perineum, and continue to the scrotum or labium. Short nerves from the pudendal or perineal nerves supply the muscles of the penis or the vestibule and vulva.

The **dorsal nerve of the penis** in the male (or of the clitoris in the female) is the main extrapelvic continuation of the pudendal nerve. It curves around the ischial arch and reaches the dorsal surface of the penis where it courses cranially. It continues through the glans penis and ends in the mucosa covering the apex of the glans. In the female, the smaller dorsal nerve of the clitoris runs ventrally to the ventral commissure of the vulva where it terminates in the clitoris.

4. The **caudal rectal nerve** may arise from sacral nerves or may leave the pudendal nerve at the caudal border of the levator ani muscle. It enters the external anal sphincter at a point slightly below its middle.

5. The **caudal cutaneous femoral nerve** (Figs. 148, 153) is united to the pudendal for most of its intrapelvic course. As it passes out of the pelvis dorsal to the ischial arch it supplies perineal branches to the skin around the anus. The caudal cutaneous femoral nerve follows the caudal gluteal artery to the level of the ischiatic tuberosity, where it becomes superficial and terminates in the skin on the proximal caudal half of the thigh.

6. The **caudal gluteal nerve** (Figs. 148, 153) passes over the greater ischiatic notch, crosses over the middle gluteal muscle and enters the medial surface of the superficial gluteal muscle. It is the sole innervation to the superficial gluteal. It has a variable origin from the seventh lumbar and first two sacral nerves.

7. The **cranial gluteal nerve** (Figs. 148, 153) passes over the greater ischiatic notch, crosses the lateral surface of the ilium at the origin of the deep gluteal muscle and innervates the middle and deep gluteal muscles and the tensor fasciae latae. It arises from the sixth and seventh lumbar and first sacral nerves.

8. The **sciatic nerve** (Figs. 155, 156) arises from the last two lumbar and first two sacral nerves. Isolate the nerve as it passes over the greater ischiatic notch. Small branches leave the trunk within the pelvis to supply the internal obturator, gemelli and quadratus femoris muscles. Do not dissect these branches. A branch leaves the nerve at the level of the hip and innervates the biceps femoris, semitendinosus, and semimembranosus muscles.

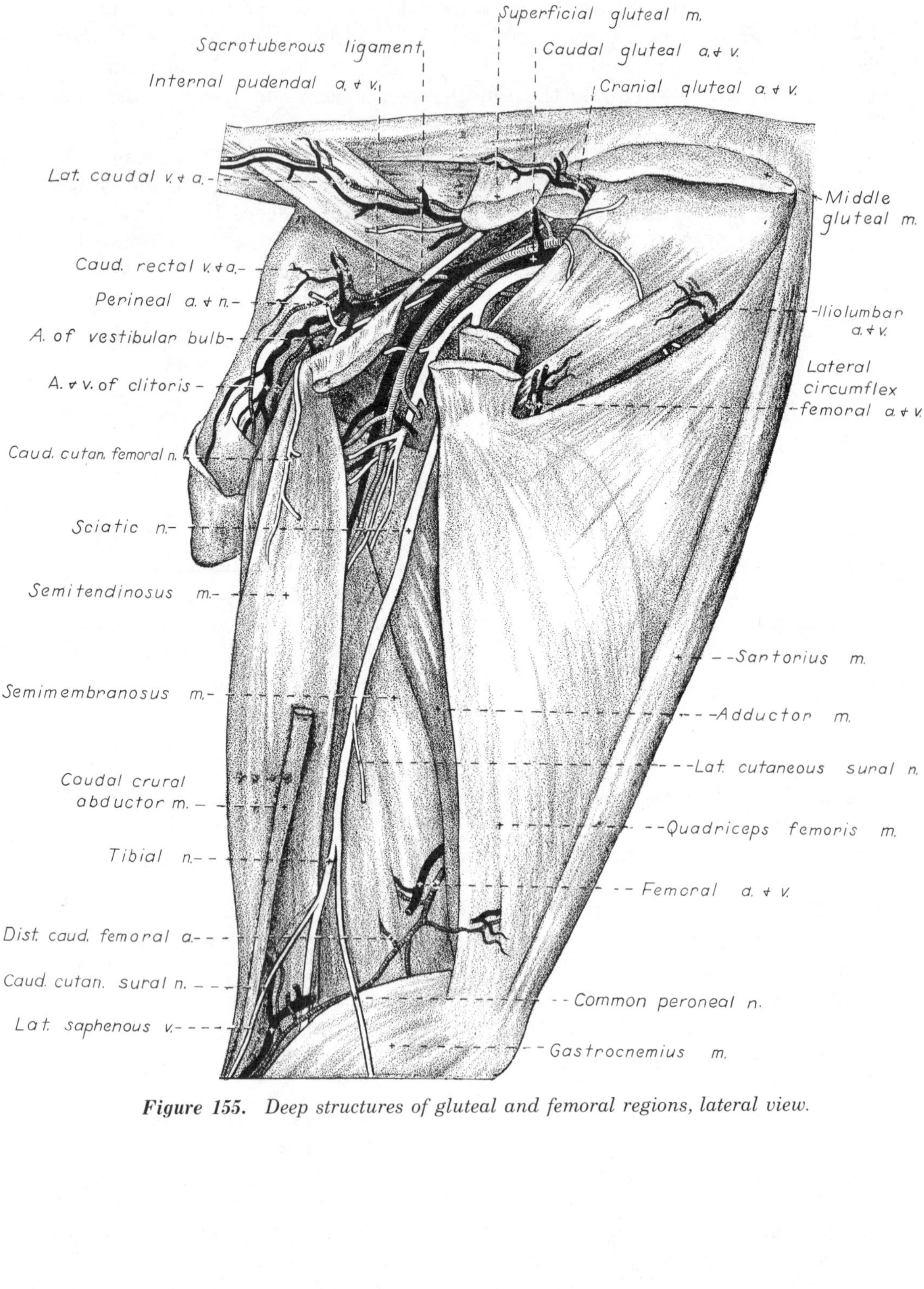

Figure 155. *Deep structures of gluteal and femoral regions, lateral view.*

There are lateral and caudal cutaneous sural nerves that arise from the ischiatic nerve in the thigh and supply the skin on the lateral and caudal surfaces of the crus. These need not be dissected.

The sciatic nerve terminates in the thigh as the common peroneal and tibial nerves. The **common peroneal nerve** (Fig. 150) is smaller and passes deep to the thin terminal part of the biceps femoris muscle. The nerve runs distally and crosses the lateral head of the gastrocnemius muscle. It passes between the lateral head of the deep digital flexor muscle caudally and the peroneus longus muscle cranially and enters the muscles on the cranial side of the crus. The common peroneal innervates the peroneus longus and divides into superficial and deep peroneal nerves.

The **superficial peroneal nerve** leaves the lateral portion of the parent nerve below the stifle where it lies between the lateral head of the deep digital flexor caudally and the peroneus longus cranially. Expose the nerve and trace it as it curves distally deep to the distal part of the peroneus longus. At the beginning of the distal third of the crus it becomes subcutaneous and joins the cranial branch of the saphenous artery. Distal to the tarsus the superficial peroneal nerve forms dorsal common digital nerves that innervate the paw. These will not be dissected.

The **deep peroneal nerve** arises from the cranial surface of the parent nerve. After entering the muscles on the cranial part of the crus with the superficial peroneal nerve, it innervates the peroneus longus, long digital extensor and cranial tibial muscles. Trace the deep peroneal nerve as it courses distally in association with the cranial tibial artery. In the proximal half of the tarsus they both lie in a groove formed by the tendons of the long digital extensor muscle laterally and by the cranial tibial muscle medially. Expose them by cutting the flexor retinaculum. At the tarsus the nerve divides into dorsal metatarsal nerves that continue distally to innervate the paw. Follow the deep peroneal nerve as the termination of the cranial tibial artery is now dissected. The terminal dorsal metatarsal nerves will not be dissected.

The cranial tibial artery is continued opposite the talocrural joint as the **dorsal pedal artery** (Fig. 157). Branches supply the tarsus, and the dorsal pedal artery terminates in the **arcuate artery.** The latter runs transversely laterally through the ligamentous tissue at the proximal end of the metatarsus. It gives off dorsal metatarsal arteries that run distally to supply the paw. These need not be dissected.

A **perforating branch** leaves dorsal metatarsal artery II, a branch of the arcuate artery, and courses distally in the space between the second and third metatarsal bones. This perforating branch passes from the dorsal to the plantar surface of the metatarsus in the proximal end of this space. It anastomoses with the caudal branch of the saphenous artery to contribute to the plantar metatarsal vessels that supply the paw. Expose the arcuate artery and the perforating branch.

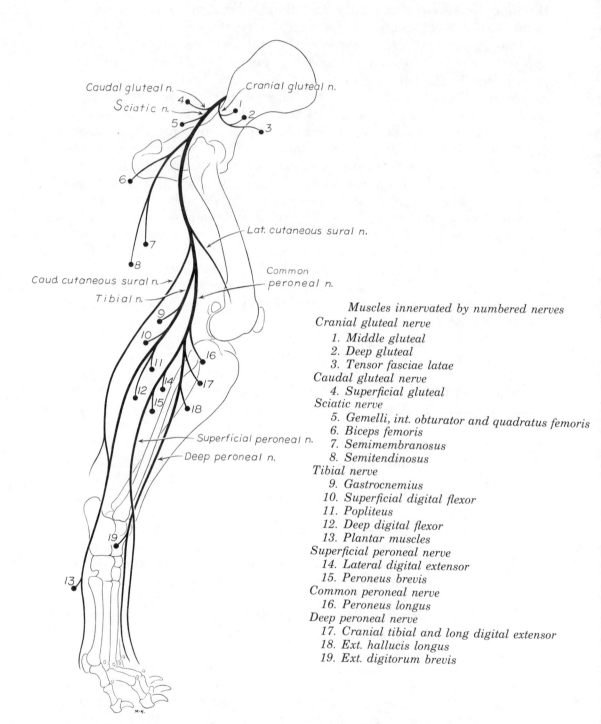

Muscles innervated by numbered nerves

Cranial gluteal nerve
1. *Middle gluteal*
2. *Deep gluteal*
3. *Tensor fasciae latae*

Caudal gluteal nerve
4. *Superficial gluteal*

Sciatic nerve
5. *Gemelli, int. obturator and quadratus femoris*
6. *Biceps femoris*
7. *Semimembranosus*
8. *Semitendinosus*

Tibial nerve
9. *Gastrocnemius*
10. *Superficial digital flexor*
11. *Popliteus*
12. *Deep digital flexor*
13. *Plantar muscles*

Superficial peroneal nerve
14. *Lateral digital extensor*
15. *Peroneus brevis*

Common peroneal nerve
16. *Peroneus longus*

Deep peroneal nerve
17. *Cranial tibial and long digital extensor*
18. *Ext. hallucis longus*
19. *Ext. digitorum brevis*

Figure 156.   Distribution of cranial and caudal gluteal nerves and sciatic nerve.

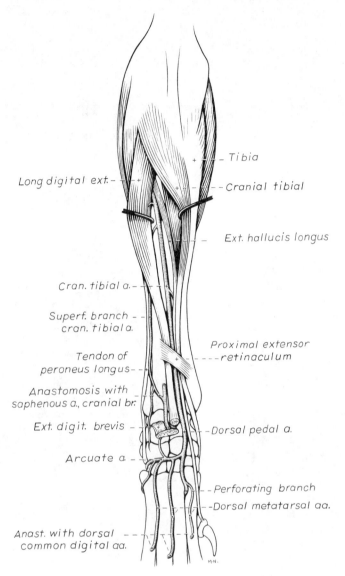

Figure 157. Cranial tibial artery, right limb.

The **tibial nerve** is the caudal portion of the sciatic nerve. It separates from the common peroneal nerve in the thigh. At the stifle it enters the crus between the two heads of the gastrocnemius muscle. The tibial nerve supplies the muscles caudal to the tibia and fibula and sends branches to the stifle and the tarsal and digital joints. It innervates both heads of the gastrocnemius muscle and the superficial digital flexor, popliteus and both heads of the deep digital flexor muscles. The tibial nerve is continued beyond these branches on the lateral head of the deep digital flexor muscle. It emerges from the deep surface of the medial head of the gastrocnemius and continues distally along the medial side of the caudal surface of the tibia. Proximal to the tarsocrural joint the tibial nerve divides into the medial and lateral plantar nerves. These cross the tarsus medial to the tuber calcis and terminate as the plantar common digital and plantar metatarsal nerves.

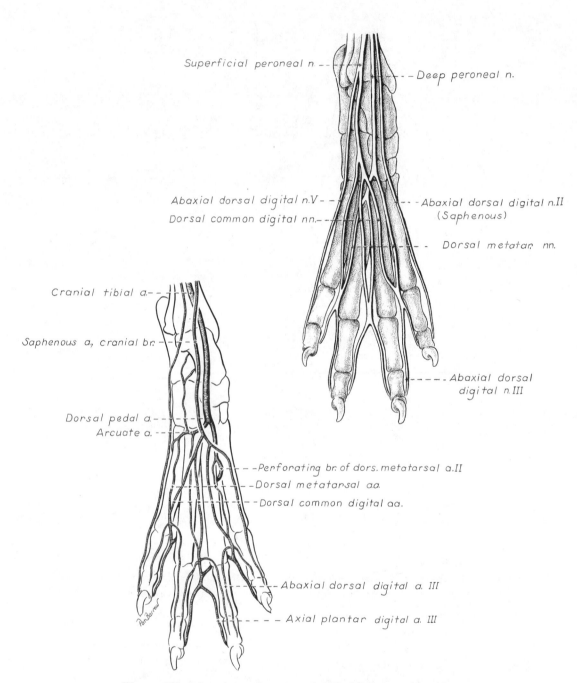

Figure 158.   Arteries and nerves of right hind paw, dorsal view.

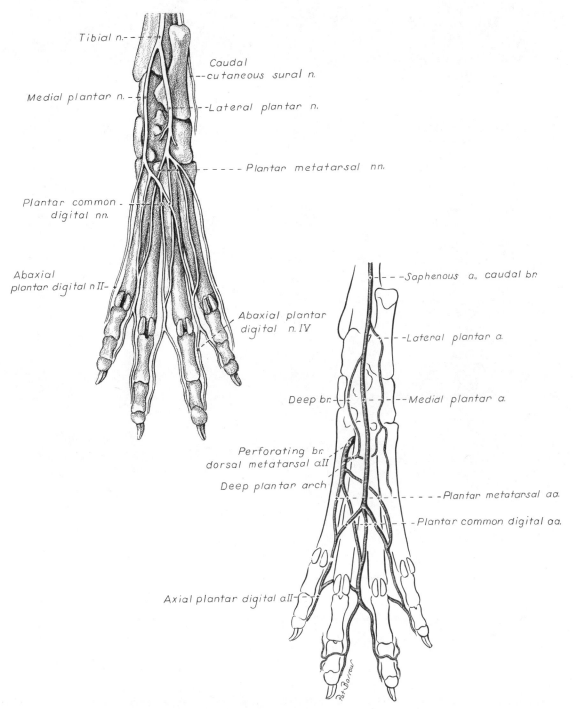

Figure 159a. *Arteries and nerves of right hind paw, plantar view.*

VESSELS AND NERVES OF THE PELVIC LIMB

MUSCLES	ARTERIAL SUPPLY	NERVE SUPPLY
Cranial Thigh Muscles		
Extensors of stifle	Lateral circumflex femoral	Femoral
Quadriceps femoris		
Medial Thigh Muscles		
Adductors of pelvic limb	Deep femorals	Obturator
Gracilis, Adductor,	Caudal femorals	
Pectineus		
Caudal Thigh Muscles		
Flexors and extensors of stifle	Deep femoral	Ischiatic
Biceps femoris,	Caudal gluteal	
Semimembranosus,	Caudal femorals	
Semitendinosus		
Cranial Muscles of Crus		
Flexors of tarsus	Cranial tibial	Peroneal
Cranial tibial,		
Peroneus longus		
Extensor of digits		
Long digital extensor		
Caudal Muscles of Crus		
Flexor of stifle	Popliteal	Tibial
Popliteus	Distal caudal femoral	
Extensor of tarsus		
Gastrocnemius		
Flexors of digits		
Superficial digital flexor		
Deep digital flexor		
Dorsal Surface of Paw		
Superficial	Saphenous	Peroneal
Deep	Dorsal pedal	
Plantar Surface of Paw		
Superficial	Saphenous	Tibial
Deep	Dorsal pedal	
	(perforating ramus)	

Blood Supply and Innervation of the Digits of the Pelvic Limb (Figs. 158–159)

DORSAL SURFACE

VESSELS

(*Superficial*)	Cranial br. saphenous	Dorsal common digitals	Axial or Abaxial dorsal
(*Deep*)	Cranial tibial Dorsal pedal – Arcuate	Dorsal metatarsals	digital a., v.

NERVES

(*Superficial*)	Superficial peroneal	Dorsal common digitals	Axial or Abaxial dorsal
(*Deep*)	Deep peroneal	Dorsal metatarsals	digital n.

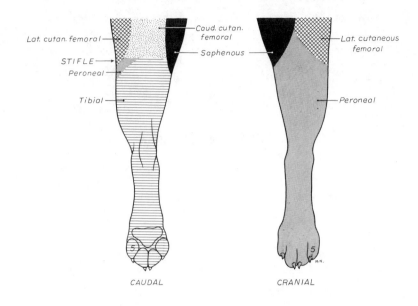

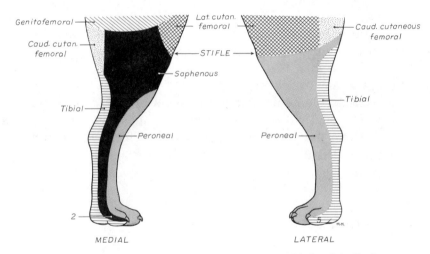

Figure 159b. *Schema of cutaneous innervation of left pelvic limb.*

PLANTAR SURFACE

VESSELS

(Superficial)	Caudal br. saphenous Medial Plantar	Plantar common digitals	Axial or Abaxial plantar
(Deep)	Deep plantar arch Perforating ramus from dorsal pedal; lateral plantar from caudal br. saphenous	Plantar metatarsals	digital a., v.

NERVES

(Superficial)	Tibial—Medial plantar	Plantar common digitals	Axial or Abaxial plantar
(Deep)	Tibial—Lateral plantar	Plantar metatarsals	digital n.

THE HEAD

THE SKULL

The **skull** or cranium is a complex of bones formed in both membrane and cartilage around the brain, sense organs and entrances to the digestive and respiratory systems. The braincase, **neurocranium,** is formed by a dermal bone roof, the **calvaria,** fused with cartilage-bone walls and floor. Articulated with the neurocranium are the bones of the face, jaws and palate, the ear ossicles and the hyoid apparatus.

Various skull elements fuse with each other developmentally or have been lost phylogenetically, with the result that each species has distinctive features that may not be present in all. The dog's skull varies more in shape among the different breeds than does the skull of other species of domestic animals.

DORSAL SURFACE OF THE SKULL (Fig. 160)

Braincase.   The paired frontal and parietal bones form the dorsum of the neurocranium. The **parietal bone** joins the frontal bone rostrally and its fellow medially. Caudally the parietal bone meets the occipital bone, which forms the caudal surface of the skull. The unpaired interparietal bone fuses with the occipital bone prenatally and appears as a rostrally extending process. The ventral border of the parietal bone joins the squamous temporal and basisphenoid bones. Rostral to the parietal bone is the **frontal bone,** which forms the dorsomedial part of the orbit.

The **sagittal crest** is a median ridge formed by the parietal and interparietal bones. It varies in height and may be absent.

In most brachycephalic breeds the sagittal crest is replaced by a pair of paramedian **temporal lines.** Each extends from the **external occipital protuberance** to the zygomatic process of the frontal bone. The **nuchal crest** is a transverse ridge which marks the transition between the dorsal and caudal surfaces of the skull. The external occipital protuberance is median in position at the caudal end of the sagittal crest.

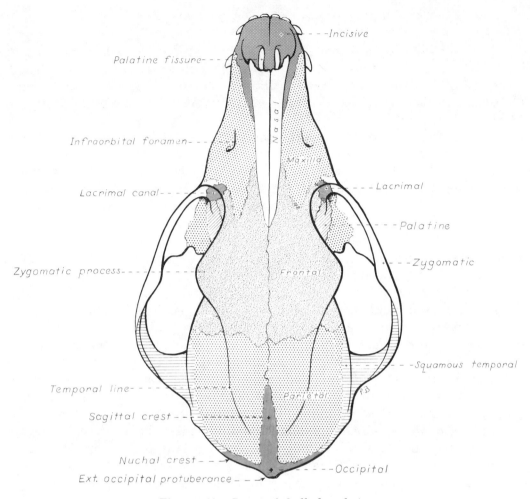

Figure 160. *Bones of skull, dorsal view.*

On each side of the dorsum of the skull is the **temporal fossa.** The floor of the fossa is convex. It is bounded medially by the sagittal crest or the temporal line, caudally by the nuchal crest, ventrally by the zygomatic process of the temporal bone and rostrally by the orbital ligament passing between the zygomatic processes of the frontal and temporal bones. The temporal fossa is continuous with the orbit. The temporal muscle arises from this temporal fossa on the frontal and parietal bones.

Facial Bones. The facial part of the dorsal surface of the skull is formed by parts of the frontal, nasal, maxillary and incisive bones. All are paired. The **nasal bone** meets its fellow on the midline, the frontal bone caudally and the maxilla and incisive bones laterally. The **maxilla** contains the upper cheek teeth. The **incisive bone** bears the three incisor teeth and has a long nasal process which articulates with the maxilla and nasal bone. The **nasal aperture** is bounded by the incisive and nasal bones. It is nearly circular in brachycephalic breeds and is oval in the dolichocephalic breeds.

LATERAL SURFACE OF THE SKULL

Braincase.   Lateral portions of the frontal and parietal bones form the lateral surface (Figs. 161 and 162) of the braincase. The caudoventral part of the lateral surface of the skull is composed largely of the **temporal bone.** This compound bone is composed of squamous, tympanic and petrous parts.

The **squamous part** forms the ventral part of the temporal fossa and bears a **zygomatic process** which forms the caudal part of the zygomatic arch. It articulates dorsally with the parietal bone, rostrally with the basisphenoid wing and caudally with the occipital bone.

The **tympanic part** of the temporal bone has a bulbous enlargement, the **tympanic bulla,** which encloses the middle ear cavity. On the lateral side of the bulla is the **external acoustic meatus.** In life the tympanic membrane closes this opening and the annular cartilage of the external ear attaches around its periphery.

The **petrosal part** of the temporal bone contains the labyrinth of the inner ear—the cochlea, the vestibule, and the semicircular canals. The major portion of this bone is visible inside the braincase. Remove the tympanic bulla on one side. This exposes a barrel-shaped eminence, the **promontory,** on the ventral surface of the petrosal bone, which contains the **cochlear window.** It is closed in life by a membrane. The **vestibular window** lies dorsal to the promontory and contains the foot-plate of the **stapes.** The stapes articulates with the **incus** which in turn articulates with the **malleus,** which is attached to the medial side of the tympanic membrane. The **mastoid process** is the only part of the petrosal bone to reach the exterior. It is small, and lies caudal to the external acoustic meatus lateral and dorsal to the root of the prominent jugular process.

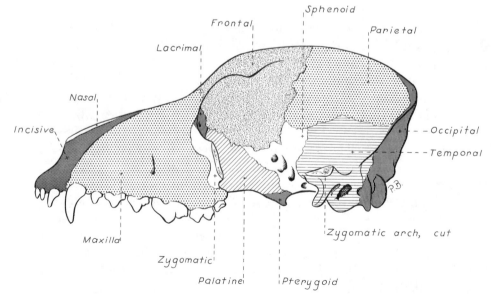

Figure 161. *Bones of skull, lateral view, zygomatic arch removed.*

The cleidomastoideus and sternomastoideus muscles insert on the mastoid process.

The **jugular process,** a part of the occipital bone, projects ventrally from a position caudal to the temporal bone. The digastricus muscle arises from this process.

Rostral to the squamous temporal bone the ventrolateral surface of the braincase is formed by the wings of the basisphenoid and presphenoid. The wing of the basisphenoid articulates caudally with the squamous part of the temporal bone, dorsally with the parietal and frontal bones and rostrally with the wing of the presphenoid.

Facial Bones. The **orbit** is the bony cavity in which the eye is situated. The **orbital margin** is formed by the frontal, lacrimal, and zygomatic bones. The lateral margin of the orbit is formed by the **orbital ligament,** which extends from the zygomatic bone to the zygomatic process of the frontal bone. The medial wall of the orbit is formed by the orbital surfaces of the frontal, lacrimal, presphenoid and palatine bones.

The **zygomatic arch** is formed by the zygomatic process of the maxilla, the zygomatic bone and the zygomatic process of the temporal bone. The arch forms the cheekbone and serves as an origin for the masseter muscle which closes the jaws.

The **pterygopalatine fossa** is located ventral to the orbit. The maxilla, palatine and zygomatic bones bound the rostral part. The caudal part is bounded by the palatine, the pterygoid and the wings of the sphenoid bones. The pterygoid muscles arise from this fossa.

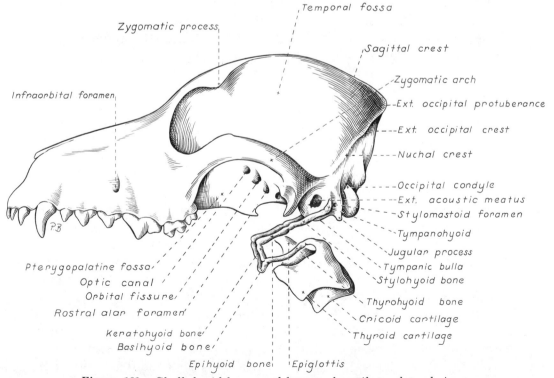

Figure 162. *Skull, hyoid bones and laryngeal cartilages, lateral view.*

The three openings in the caudal part of the orbit are, from rostral to caudal, the **optic canal,** the **orbital fissure** and the **rostral alar foramen.** The optic canal passes through the presphenoid and the rostral alar foramen passes through the basisphenoid. The orbital fissure is formed in the articulation between the basisphenoid and presphenoid bones.

In the rostral part of the pterygopalatine fossa are several foramina. The **caudal palatine foramen** and **sphenopalatine foramen** are closely related openings of about equal size located on the rostromedial wall of the fossa. The sphenopalatine foramen is dorsal to the caudal palatine. The major palatine artery, vein, and nerve enter the palatine canal through the caudal palatine foramen and together course to the hard palate. The sphenopalatine artery and vein and the caudal nasal nerve enter the nasal cavity via the sphenopalatine foramen. Rostrolateral to these is the **maxillary foramen,** the caudal opening of the infraorbital canal. The infraorbital artery, vein, and nerve course rostrally through this canal. A small part of the rostromedial wall of the pterygopalatine fossa, just caudal to the maxillary foramen, often presents an opening which is normally occupied by a thin plate of bone, which serves as the origin of the ventral oblique eye muscle. Caudal to the maxillary foramen are a number of small openings, most of them for the small nerves and vessels which pass through their respective **alveolar canals** to the roots of the last two cheek teeth and the caudal root of the last premolar. Above the maxillary foramen in the lacrimal bone is the shallow **fossa for the lacrimal sac.** The fossa is continued by the nasolacrimal canal for the nasolacrimal duct.

The facial part of the lateral surface of the skull rostral to the orbit includes the lateral surface of the maxilla and the incisive bone. Dorsal to the third premolar tooth is the **infraorbital foramen,** the rostral opening of the infraorbital canal. The roots of the cheek teeth produce lateral elevations, the **alveolar juga.**

Ventral Surface of the Skull (Figs. 163, 164)

Braincase. The basal portion of the occipital bone forms the caudal third of the cranial base. It articulates laterally with the tympanic and petrous parts of the temporal bone and rostrally with the body of the basisphenoid. Caudally the occipital condyle articulates with the atlas. The jugular process, a projection of the occipital bone, articulates with the caudolateral part of the tympanic bulla. The basisphenoid articulates caudally with the occipital and rostrally with the presphenoid and pterygoid bones. The oval foramen and the alar canal pass through the basisphenoid bone. The presphenoid articulates caudally with the basisphenoid and pterygoid, laterally with the perpendicular part of the palatine and rostrally with the vomer. Only a small median portion of the presphenoid is exposed on the ventral surface of the braincase. The **rostral alar foramen,** caudoventral to the orbital fissure, is the rostral opening of the **alar canal.** The caudal opening of this short canal is the **caudal alar foramen.**

The **round foramen** opens from the cranial cavity into the alar canal. The maxillary nerve from the trigeminal nerve enters the alar canal from the neurocranium via this foramen. The nerve courses rostrally and leaves the alar canal via the rostral alar foramen. In addition the maxillary artery traverses the full length of the alar canal.

The **oval foramen,** a direct opening into the cranial cavity, is caudolateral to the caudal alar foramen. The mandibular nerve from the trigeminal nerve leaves the cranial cavity through this opening.

The **foramen lacerum** lies at the rostromedial edge of the tympanic bulla. A loop of the internal carotid artery protrudes through this opening.

The **musculotubal canal** lies lateral to the foramen lacerum and caudal to the oval foramen. It is the bony enclosure of a tubular connection, the auditory tube, which runs from the middle ear to the pharynx.

The **tympanooccipital fissure** is an oblong depression between the basilar part of the occipital bone and the tympanic part of the temporal bone. The petro-occipital canal and the carotid canal leave the depths of the fissure at about the same place. The carotid canal transmits the

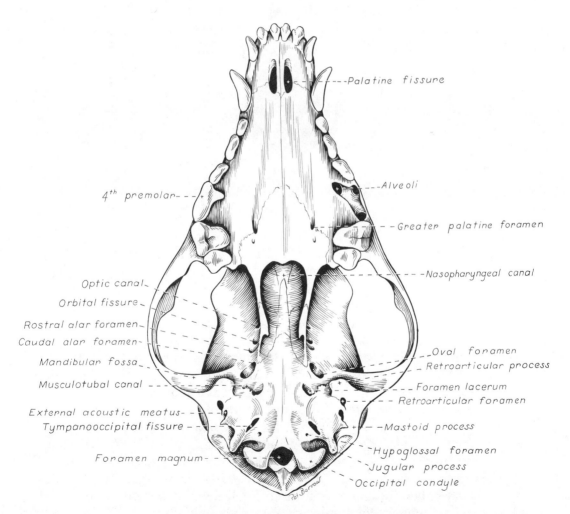

Figure 163. Skull, ventral view.

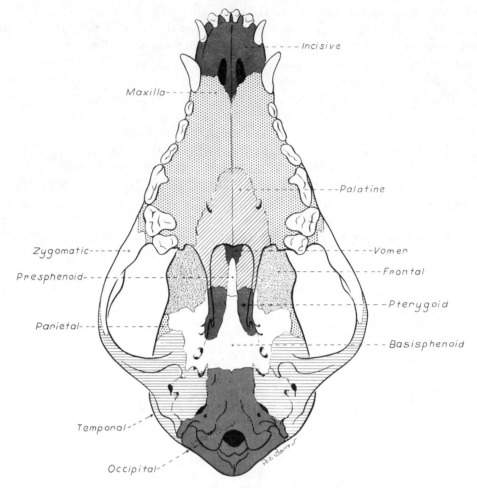

Figure 164. Bones of skull, ventral view.

internal carotid artery. The petro-occipital canal transmits a vein. Neither canal can be adequately demonstrated on an articulated skull. The glossopharyngeal, vagus and accessory nerves course peripherally through this fissure.

The **hypoglossal canal,** for the passage of the hypoglossal nerve, lies caudal to the petro-occipital fissure in the occipital bone.

The **mandibular fossa** of the temporal bone articulates with the condyle of the mandible to form the temporomandibular joint. The **retroarticular process** forms the caudal wall of the mandibular fossa.

Between the tympanic bulla and the mastoid process of the temporal bone is the **stylomastoid foramen.** This is the opening of the facial canal which conducts the facial nerve peripherally through the petrosal bone.

Facial Bones.   The ventral surface of the facial part of the skull is characterized by the teeth and hard palate. There is an upper (maxillary) and a lower (mandibular) dental arch. Each tooth lies in an alveolus or socket. An alveolus is subdivided by interradicular septae for those teeth with more than one root. The **hard palate** is composed of the horizontal parts of the palatine, maxillary and the incisive bones. A pair of openings, the **palatine fissures,** is located on each side of the midline between the canine teeth.

The palatine bones form the caudal third of the hard palate. The **major palatine foramen** is medial to the fourth cheek tooth. Caudal to this is the **minor palatine foramen.** The major palatine artery, vein and nerve, and their branches, emerge through these foramina. The **choanae** are the openings of the right and left nasal cavities into the nasopharynx. They are located at the caudal end of the hard palate.

Caudal Surface of the Skull

Developmentally, the occipital bone is formed by paired exoccipitals (which bear the condyles), a supraoccipital and a basioccipital. The lateral borders form a **nuchal crest** and, middorsally, an **external occipital protuberance.** The **foramen magnum** is the large opening into the cranial cavity through which the spinal cord is continued as the brain stem. The **mastoid foramen** is located in the occipitotemporal suture, dorsolateral to the occipital condyle. It transmits the caudal meningeal artery and vein. The rest of the caudal surface of the skull is roughened for muscular attachment.

Mandible

The mandible (Fig. 165), or lower jaw, bears the lower teeth and articulates with the temporal bone. The two mandibles join rostrally at the symphysis. Each mandible can be divided into a **body** or horizontal part and a **ramus** or perpendicular part. The **alveolar border** of the mandible contains alveoli for the roots of the teeth.

On the lateral surface of the ramus of the mandible is the triangular **masseteric fossa** for the insertion of the masseter muscle. The dorsal half of the ramus is the **coronoid process.** Its medial surface has a shallow depression for insertion of the temporal muscle. Ventral to this is the **mandibular foramen.** This foramen is the caudal opening of the **mandibular canal,** which is located in the ramus and body of the mandible. It transmits the mandibular artery and vein and the mandibular alveolar nerve. It opens rostrally at the three mental foramina. The pterygoid muscles insert on the medial surface of the mandible and on the angular process, ventral to the insertion of the temporal muscle.

The **condyloid process** enters into the formation of the temporomandibular joint. Between the condyloid process and the coronoid process is a U-shaped depression, the **mandibular notch.**

The **angular process** is a hooked eminence ventral to the condyloid process. The digastricus muscle inserts on this process.

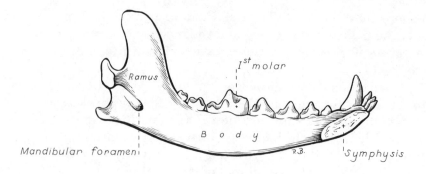

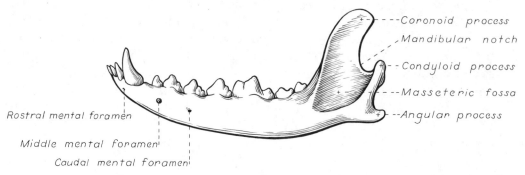

Figure 165. *Mandible, medial view and lateral view.*

HYOID BONES

The **hyoid apparatus** (Fig. 166) is composed of the hyoid bones, which stabilize the tongue and the larynx. This apparatus extends from the mastoid process to the thyroid cartilage of the larynx and consists of the tympanohyoid cartilage and the following articulated bones: the stylohyoid, epihyoid, keratohyoid, basihyoid and thyrohyoid. All of the bones are paired except for the basihyoid, which unites the elements of the two sides in the root of the tongue. Examine the wet specimens provided as well as your own specimen.

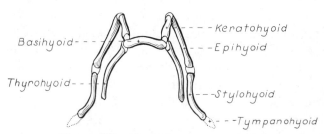

Figure 166. *Hyoid bones, ventral view.*

Teeth

The teeth are arranged in maxillary and mandibular arches which face each other. The mandibular arch is narrower than the maxillary.

The upper teeth are contained in the incisive and maxillary bones. Those whose roots are embedded in the incisive bone are known as the **incisor teeth.** Caudal to these and separated from them by a space is the **canine tooth.** Behind this are the **cheek teeth,** which are divided into **premolars** and **molars.** The lower teeth are in general similar to the upper. There is one more molar tooth in each mandible than in the corresponding maxilla. Some of the teeth—the incisors, the fourth premolar and the molars—usually meet those of the opposite arch when the jaws are closed. The first three premolars fail to meet during normal closure and the opening between the teeth is called the premolar carrying space.

The dental formula for the permanent teeth of the dog is

$$\text{I} \quad \text{C} \quad \text{Pm} \quad \text{M}$$

$$\begin{array}{c} Upper \\ Lower \end{array} \quad \begin{array}{c} 3 \\ 3 \end{array} \quad \begin{array}{c} 1 \\ 1 \end{array} \quad \begin{array}{c} 4 \\ 4 \end{array} \quad \frac{2}{3} = \frac{10}{11} \times 2 = 42$$

The temporary or deciduous dentition ("milk teeth") can be expressed by the formula

$$\text{I} \quad \text{C} \quad \text{Pm}$$

$$\begin{array}{c} Upper \\ Lower \end{array} \quad \begin{array}{c} 3 \\ 3 \end{array} \quad \begin{array}{c} 1 \\ 1 \end{array} \quad \frac{3}{3} = \frac{7}{7} \times 2 = 28$$

The first incisor tooth (central) is next to the median plane, and is followed by the second incisor (intermediate) and the third incisor (corner). In the permanent dentition the premolar and molar teeth are numbered from rostral to caudal; thus the tooth nearest the canine is number one. The fourth premolar is the largest cheek tooth of the maxilla; the largest cheek tooth of the mandible is the first molar. These are known as the sectorial or shearing teeth.

The first premolar in the dog has no deciduous predecessor. The teeth of the permanent set are much larger than those of the deciduous set. The last permanent tooth erupts at six or seven months.

Each tooth possesses a **crown** and a **root** (or roots) which are embedded in the alveoli of the jaws. The junction of root and crown is the **neck** of the tooth.

The roots of the teeth are fairly constant, the incisors and canines of both jaws having one each. In the upper jaw the first premolar has one root; the second and third have two each; the fourth premolar and the first and second molars have three each. The lower cheek teeth have two roots each, except the first premolar and third molar, which have one. The **upper shearing tooth,** which is the fourth premolar, has two rostral roots in a transverse plane and a large caudal root. Notice that the lateral roots of the fourth premolar lie ventrolateral to the infraorbital canal and that the medial root of the rostral pair lies ventromedial to the infraorbital foramen.

The outer surface of the teeth is the **vestibular surface** and the inner surface is the **lingual surface.** The sides of a tooth which lie in contact with or face an adjacent tooth are the **contact surfaces.** The surface of the tooth which faces the opposite dental arch is known as the **occlusal** or **masticating surface.**

CAVITIES OF THE SKULL

Cranial Cavity.   The cranial cavity (Figs. 167, 168) contains the brain and its coverings and vessels. The roof of the skull, the calvaria, is formed by the parietal and frontal bones. The rostral two-thirds of the base of the cranium is formed by the sphenoid and occipital bones. The caudal wall is formed by the occipital bone, and the rostral wall by the cribriform plate of the ethmoid bone. The lateral walls are formed by the temporal, parietal, frontal and sphenoid bones. The interior of the cranial cavity contains

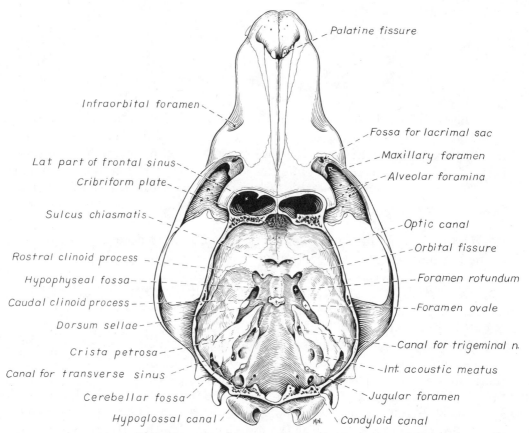

Figure 167. Skull with calvaria removed, dorsal view.

impressions formed by the gyri and sulci of the brain. Arteries leave grooves on the cerebral surface of the bones, while many of the veins lie in the diploe between the outer and inner tables of the bones. The base of the cranial cavity is divided into rostral, middle and caudal fossae.

The **rostral fossa** is located in front of the optic canals. The floor of this fossa is formed by the presphenoid bone and the cribriform plate of the ethmoid. It is bounded laterally by the frontal bone. The olfactory bulbs and the rostral parts of the frontal lobes of the brain lie in this fossa. The numerous **cribriform foramina** in the cribriform plate transmit olfactory nerves from the olfactory epithelium of the nasal cavity to the olfactory bulbs of the brain. The **optic canal** is a short passage in the presphenoid bone through which the optic nerves course.

The **middle cranial fossa** extends caudally from the optic canals to the petrosal crests and the dorsum sellae. It is situated on a lower level than the rostral cranial fossa. Several paired foramina are found on the floor of this fossa. Caudal to the optic canal is the orbital fissure. The oculomotor, trochlear and abducens nerves and the ophthalmic nerve from the trigeminal nerve leave the cranial cavity through this opening. Caudal and lateral to the orbital fissure is the round foramen that transmits the maxillary nerve from the trigeminal nerve to the alar canal. Caudolateral to the round foramen is the oval foramen that transmits the mandibular nerve from the trigeminal nerve and the middle meningeal artery.

The **sella turcica** of the basisphenoid contains the hypophysis. It is composed of a shallow **hypophyseal fossa,** limited rostrally by the presphenoid bone and bounded caudally by a raised quadrilateral process, the **dorsum sellae.** The caudal part of the middle fossa is the widest part of the cranial cavity. The parietal and temporal lobes of the cerebrum are located here.

The **caudal cranial fossa** contains the cerebellum, the pons and the medulla, and part of the occipital lobes of the cerebrum. It extends from the petrosal crests and dorsum sellae to the foramen magnum. The floor of this fossa, except for a small part at the rostral end, is formed by the basioccipital bone.

Study the foramina and canals on the floor and sides of the cranial fossa (Fig. 168). The opening of the carotid canal is located under the rostral tip of the petrosal bone.

The **canal for the trigeminal nerve** is in the rostral end of the petrosal bone. The trigeminal ganglion is located in the canal. Caudal to this is the **internal acoustic meatus,** through which the facial and vestibulocochlear nerves pass. Dorsocaudal to the internal acoustic meatus is the **cerebellar fossa** that contains a small lateral portion of the cerebellum.

The **jugular foramen** is located between the petrosal and the occipital bones. It opens to the outside through the occipitotympanic fissure and transmits the glossopharyngeal, vagus and accessory cranial nerves. Caudomedial to the jugular foramen is the hypoglossal canal. Dorsal to this foramen is the **condyloid canal,** which transmits a venous sinus.

Projecting rostroventrally from the caudal wall of the cranial cavity

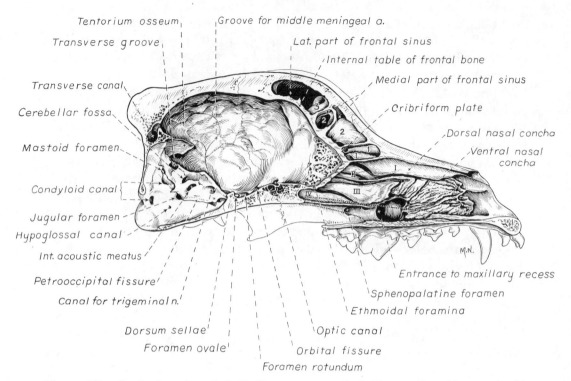

Figure 168. *Sagittal section of skull. (Roman numerals indicate endoturbinates; Arabic numerals indicate ectoturbinates.)*

is the **tentorium osseum.** This is composed of processes from the parietal and occipital bones. The dural membrane, the **tentorium cerebelli,** attaches to the petrosal crests and the tentorium osseum, separating the cerebrum from the cerebellum. A relatively median foramen for the dorsal sagittal sinus is located on the rostral surface of the occipital bone dorsal to the tentorium osseum. It opens into the paired **transverse canals.** This foramen transmits the dorsal sagittal venous sinus to the transverse sinus in the transverse canal. The transverse sinus continues ventrolaterally through the transverse groove of the occipital bone and then as the temporal sinus passes through the temporal bone lateral to the petrous portion. At the retroarticular foramen the temporal sinus communicates with the maxillary vein.

Nasal Cavity. The nasal cavity is the facial part of the respiratory passage. It begins at the bony **nasal aperture** and is composed of two symmetrical halves separated by a median nasal septum. The caudal openings of the nasal cavities are the choanae. In a median-sectioned skull study the bony scrolls, the conchae, which lie in the nasal fossa. Compare them with the mucosa-covered conchae of a hemisected embalmed head.

The conchae (Figs. 168, 169) project into each nasal fossa and, with their mucosa, act as baffels to warm and cleanse inspired air. Their caudal portions also serve as locations of olfactory nerve cells, whose axons course to the olfactory bulbs of the brain through the cribriform plate.

The **dorsal nasal concha** originates as the most dorsal scroll on the cribriform plate and extends rostrally as a shelf attached along the medial surface of the nasal bone.

The **ventral nasal concha** consists of several elongate scrolls which attach to a crest on the medial surface of the maxilla. It lies in the middle of the nasal cavity but does not contact the median nasal septum.

The **ethmoidal labyrinth** is composed of many delicate scrolls that attach to the cribriform plate and occupy the fundus of the nasal cavity. Dorsally the scrolls extend as ectoturbinates into the rostral portion of the frontal sinus. Ventrally, as endoturbinates, the scrolls attach to the vomer, which separates the entire ethmoidal labyrinth from the nasopharynx.

The **ethmoid bone** complex is located between the braincase and the facial part of the skull. It consists of the ethmoidal labyrinth, the cribriform plate and the median bony perpendicular plate of the nasal septum. This complex is surrounded by the frontal bone dorsally, the maxilla laterally and the vomer and the palatine ventrally.

The **nasal septum** separates the right and left nasal cavities. It is composed of cartilage and bone. The cartilaginous part, the **septal cartilage,** forms the rostral two-thirds of this median partition. It articulates with other cartilages at the nares, which prevent collapse of the nostrils. Ventrally the septal cartilage fits into a groove formed by the vomer and dorsally it articulates with the nasal bones where they meet at the midline. The osseous part of the nasal septum is formed by the perpendicular plate of the ethmoid bone, the septal processes of the frontal and nasal bones and the sagittal portion of the vomer.

In each nasal cavity the shelf-like dorsal nasal concha and the scrolls of the ventral nasal concha divide the cavity into four primary passages known as meatuses. The **dorsal nasal meatus** lies between the nasal bone and the dorsal nasal concha. The small **middle nasal meatus** lies between the dorsal nasal concha and the ventral nasal concha, while the **ventral**

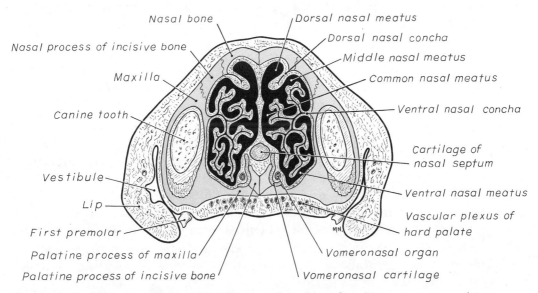

Figure 169. *Cross-section of nasal cavity.*

nasal meatus is dorsal to the hard palate. Since the conchae do not reach the nasal septum a vertical space, the **common nasal meatus,** is formed on each side of the nasal septum. This space extends from the nasal aperture to the choanae in a longitudinal direction and from the nasal bone to the hard palate in a vertical direction.

Paranasal Sinuses.   The **maxillary recess,** not a true sinus, communicates with the nasal cavity. Its opening lies in a transverse plane through the rostral roots of the upper fourth premolar tooth. The recess continues caudally to a plane through the last molar tooth. The walls of the maxillary recess are formed laterally and ventrally by the maxilla and medially by the external lamina of the ethmoid bone.

There are three frontal sinuses (Fig. 168) located between the outer and inner tables of the frontal bone, divided into lateral, rostral and medial. The **lateral frontal sinus,** the largest, occupies the zygomatic process. It may be partly divided by osseous septa extending into it. The **rostral** and **medial sinuses** are irregular. The ethmoid labyrinth bulges into the medial sinus. All three sinuses communicate with the nasal cavity.

Tympanic Cavity.   The tympanic cavity is the cavity of the middle ear. It houses the auditory ossicles and communicates with the nasopharynx via the auditory tube. It is bounded ventrally by the tympanic bulla and dorsally by the petrosal bone. Laterally the external acoustic meatus is closed by the tympanic membrane.

STRUCTURES OF THE HEAD

In order to facilitate the dissection of the head it should be removed and divided on the median plane. Transect all structures ventral to the fourth cervical vertebra. This will include the trachea and esophagus and associated vessels and nerves. Reflect these from the neck to the level of the atlanto-occipital joint. Flex the head, sever the epaxial muscles over the cranial end of the atlas and expose the atlanto-occipital membrane that covers the spinal cord. Sever this membrane and the underlying spinal cord and disarticulate the atlanto-occipital joint by cutting its ligamentous attachment.

Cut the head on the median plane on the band saw and wash the cut surface to remove bone dust and hair.

Skin both halves, leaving the muscles in place. Leave a narrow rim of skin around the margin of the eyelids and at the edge of the lips. Preserve the nose and the **philtrum,** which is the median groove separating the right and left parts of the upper lip. Skin only the base of the ear.

The right side of the head will be used for the dissection of vessels and nerves, while the left side will be used for muscles. At the end of each dissection period wrap the head in cheesecloth moistened with phenol water or Phenoxetol before covering your specimen. A plastic bag is useful to prevent desiccation.

MUSCLES OF THE FACE

The muscles of the face function to open, close or move the lips, eyelids, nose and ear.

Cheek, Lips and Nose. The **platysma** (Fig. 170) is a cutaneous muscle that passes from the middorsal raphe of the neck to the angle of the mouth, where it radiates into the orbicularis oris. It is the most superficial muscle covering the ventrolateral surface of the face. Transect this muscle and reflect both ends.

The **orbicularis oris** lies near the free borders of the lips and extends from one lip to the other around the angle of the mouth. The fibers of each side end at the midline in the incisor region of both jaws.

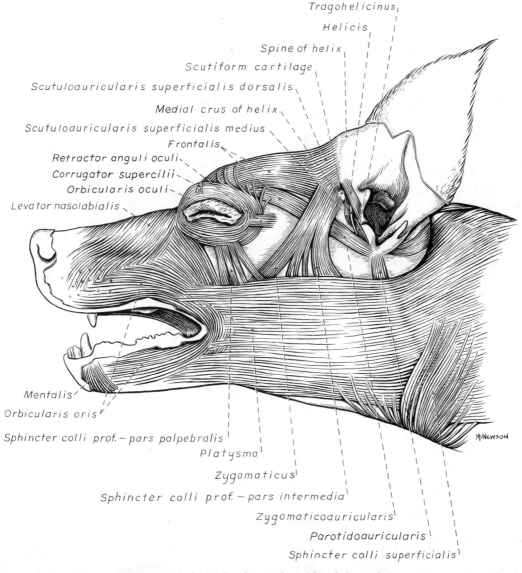

Figure 170. *Superficial muscles of head, lateral view.*

The **buccinator muscle** is a wide muscle which forms the foundation of the cheek. It attaches to the alveolar margins of the mandible and maxilla and the adjacent buccal mucosa. It may be found between the rostral border of the masseter muscle and the orbicularis oris. A portion of the buccinator lies deep to the orbicularis oris muscle and is difficult to separate from it. Its action is to return food from the vestibule to the occlusal surface of the teeth.

The **levator nasolabialis** is a flat muscle lying beneath the skin on the lateral surface of the nasal and maxillary bones. It arises from the maxillary bone, courses rostroventrally and attaches to the edge of the upper lip and on the external naris. It dilates the nostril and raises the upper lip.

Eyelids. Before dissecting the eyelids, the **palpebrae,** observe their external features.

The upper eyelid bears cilia (eyelashes) on its free border. The lower eyelid lacks cilia. The cutaneous or external surface of the eyelid is covered by hair. The posterior or inner surface is covered by a mucous membrane, the **palpebral conjunctiva.** The palpebral conjunctiva is reflected from the eyelids onto the globe of the eye as the **bulbar conjunctiva.** The angle formed by this reflection is called the **fornix.** The potential cavity thus formed, the **conjunctival sac,** is bounded posteriorly by the bulbar conjunctiva and the cornea and anteriorly by the palpebral conjunctiva.

The two lids unite laterally and medially to form the **medial** and **lateral palpebral commissures.** The **lacrimal punctum** of each lid is the beginning of the dorsal and ventral lacrimal ducts. Each is a small opening on the conjunctival margin of the lid a few millimeters from the medial commissure. The **lacrimal gland,** located ventral to the zygomatic process of the frontal bone, secretes into the dorsolateral part of the conjunctival sac. After this serous fluid has passed across the cornea it is collected by the puncta, and passes in succession through the **lacrimal duct** of each lid, the **lacrimal sac** and the **nasolacrimal duct** to the ventral nasal meatus of the nasal cavity, where evaporation of the lacrimal secretion takes place. The **caruncula lacrimalis** is a small, pyramidal body, usually pigmented, which lies in the medial palpebral commissure. The lacrimal gland and the rostral opening of the nasolacrimal duct will be seen later. The opening between the lids is the **palpebral fissure.**

The **plica semilunaris** or **third eyelid** (see Fig. 180) is a concave shelf of conjunctiva and cartilage which protrudes laterally from the medial commissure of the eye. The cartilage extends into the orbit and is surrounded by a body of fat and glandular tissue, the **superficial gland of the third eyelid.** This will be dissected shortly. Its secretion enters the conjunctival sac under the third eyelid at the medial commissure. Lift the third eyelid from the bulbar conjunctiva and examine its medial surface. Note the slightly raised lymphoid tissue.

There are several muscles associated with the eyelids (Fig. 170).

The **orbicularis oculi** lies partly in the eyelids and is attached medially to the medial palpebral ligament. Laterally the fibers of the muscle blend with those of the retractor anguli oculi. The action of the muscle is to close the palpebral fissure. The **levator palpebrae superioris** arises deep within the orbit and will be dissected with the muscles of the eyeball. It elevates the upper lid.

The External Ear. The **rostral auricular muscles** include those muscles which lie on the forehead caudal to the orbit and which converge toward the auricular cartilage. Transect the muscles at their origins and turn them toward the auricular cartilage. Notice that the middorsal part arises from its fellow of the opposite side.

The **scutiform cartilage** is a small, boot-shaped, cartilaginous plate located in the muscles rostral and medial to the external ear. It is an isolated cartilage interposed in the auricular muscles.

The **caudal auricular muscles** are the largest group. Most of these muscles arise from the median raphe of the neck and attach directly to the auricular cartilage. Transect the caudal auricular muscles and turn the external ear ventrally to expose the temporal muscle.

ORAL CAVITY

The oral cavity or mouth is divided into the vestibule and the oral cavity proper. The **vestibule** is the cavity lying outside the teeth and gums and inside the lips and cheeks. The ducts of the parotid and zygomatic salivary glands open into the dorsocaudal part of the vestibule. The **parotid duct** opens through the cheek on a small papilla located opposite the caudal end of the upper fourth premolar or shearing tooth. The **major duct of the zygomatic gland** opens lateral to the caudal part of the upper first molar tooth.

The **oral cavity proper** is bounded dorsally by the hard palate and a small part of the adjacent soft palate; laterally and rostrally by the dental arches; ventrally by the tongue and adjacent mucosa. It freely communicates with the vestibule by numerous interdental spaces and is continued caudally by the oropharynx.

Examine the tongue. It is a muscular organ composed of the interwoven bundles of intrinsic and extrinsic muscles. These will be dissected later. It is divided into a **root,** which composes its caudal third; a **body,** which is the long slender rostral part of the tongue; and a free extremity, the **apex.** The mucosa covering the dorsum of the tongue is modified to form various types of papillae. A dissecting microscope facilitates their examination. Five types of papillae are recognized by their shape. The **filiform papillae** are found predominately on the rostral two-thirds of the tongue. They are arranged in rows like shingles with their multiple pointed tips directed caudally. At the root of the tongue the filiform papillae are replaced by **conical papillae,** which have only one pointed tip. The **fungiform papillae** have a smooth, rounded surface and are located among the filiform papillae. A few may be scattered caudally among the conical papillae. The **foliate papillae** are found on the lateral margins of the root of the tongue, rostral to the palatoglossal arch. They are leaf-like, but appear as a row of parallel grooves in the fixed specimen. The **vallate papillae** are located at the junction of the body and root of the tongue. There are four to six in the dog, arranged in the form of a V with the apex directed caudally. They are larger than the others and have a circular surface, and they are surrounded by a sulcus.

The tongue is attached rostrally to the floor of the oral cavity by a ventral median fold of mucosa, the **lingual frenulum.** Examine the medial cut surface of the apex of the tongue. On the midline, just under the mucosa, is the **lyssa.** This fusiform fibrous spicule extends from the apex to the level of the attachment of the frenulum. Expose the lyssa.

Turn the tongue medially and notice that closely adjacent to the frenulum laterally and protruding from the floor of the oral cavity is a slightly raised elevation. This is the **sublingual caruncle.** Extending caudally from the caruncle is the **sublingual fold.** The **mandibular duct** and **major sublingual duct** (Fig. 171) are found in this fold. They course rostrally to open on or beside the sublingual caruncle, separately or through a common opening. The major sublingual duct is connected caudally with the monostomatic **sublingual gland.** This is closely associated with the mandibular salivary gland. There are also sublingual gland lobules (the polystomatic sublingual gland) deep to the mucosa of the sublingual fold, which have independent microscopic ducts opening into the oral cavity. Cut through the sublingual fold beside the frenulum and the body of the tongue. Notice the ducts and the salivary tissue deep to the mucosa. Follow the mandibular and major sublingual ducts to the root of the tongue where they arise from their respective salivary glands.

The **mandibular salivary gland** (Fig. 171) lies between the maxillary and linguofacial veins. It is more thickly encapsulated than the parotid. Make a longitudinal incision through its capsule and free the caudal part of the gland. The duct leaves the medial surface of the gland near its rostral end. Here it is surrounded by the sublingual gland, which has its own duct. Initially the two ducts are related to the masseter muscle and the mandible laterally and the digastricus muscle medially. They course over the digastricus to reach a position between the styloglossus medially and the mylohyoideus laterally.

The **parotid salivary gland** lies between the mandibular gland and the ear. It is closely applied to the base of the ear. The **parotid duct** is formed by two or three converging radicles which unite and leave the rostral border of the gland. It grooves the lateral surface of the masseter muscle as it passes to the cheek. It opens into the vestibule on a small papilla at the level of the caudal margin of the fourth upper premolar. Evert the upper lip near the commissure and find the small opening into the vestibule.

Examine the roof of the oral cavity. The hard palate is crossed by approximately eight transverse ridges. A small eminence, the **incisive papilla,** is located just caudal to the central incisor teeth. The fissure on

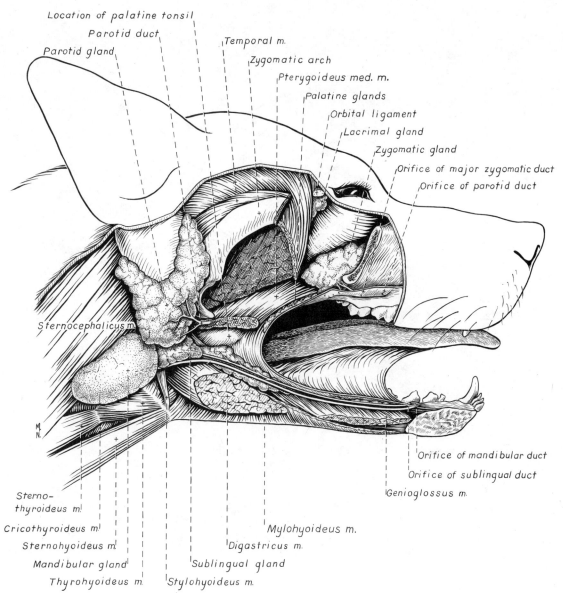

Location of palatine tonsil
Parotid duct
Parotid gland
Temporal m.
Zygomatic arch
Pterygoideus med. m.
Palatine glands
Orbital ligament
Lacrimal gland
Zygomatic gland
Orifice of major zygomatic duct
Orifice of parotid duct
Sternocephalicus m.
M. N.
Sterno-thyroideus m.
Cricothyroideus m.
Sternohyoideus m.
Mandibular gland
Thyrohyoideus m.
Stylohyoideus m.
Sublingual gland
Digastricus m.
Mylohyoideus m.
Genioglossus m.
Orifice of sublingual duct
Orifice of mandibular duct

Figure 171. Salivary glands.

either side of this papilla contains the oral opening of the **incisive duct.** Its course through the palatine fissure and the nasal opening in the ventral nasal meatus will not be dissected. The **vomeronasal organ,** an elongate structure at the base of the nasal septum, communicates with this duct. It is olfactory in nature and need not be dissected.

PHARYNX

The pharynx (Figs. 172, 173) is a passage which is, in part, common to both the respiratory and digestive systems. It is divided into oral, nasal and laryngeal parts. The **oropharynx** extends from the level of the palatoglossal arches to the caudal border of the soft palate and the base of the

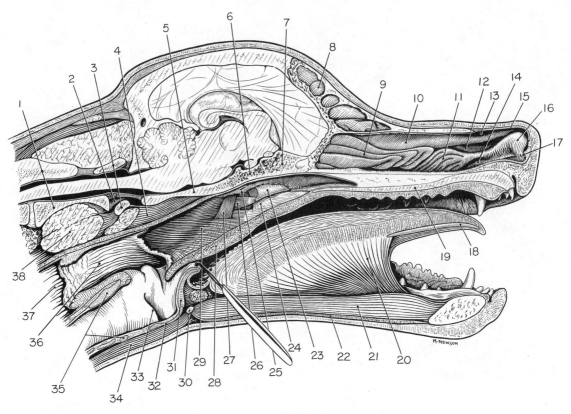

Figure 172. *Sagittal section of head.*

1. *Axis*	20. *Genioglossus*
2. *Dens*	21. *Geniohyoideus*
3. *Atlas*	22. *Mylohyoideus*
4. *Longus capitis*	23. *Pterygoid bone*
5. *Basioccipital*	24. *Tensor veli palatini*
6. *Basisphenoid*	25. *Pharyngeal orifice of auditory tube*
7. *Presphenoid*	26. *Pterygopharyngeus*
8. *Frontal sinus*	27. *Levator veli palatini*
9. *Ethmoid labyrinth*	28. *Soft palate*
10. *Dorsal nasal concha*	29. *Palatopharyngeus*
11. *Ventral nasal concha*	30. *Basihyoid*
12. *Middle nasal meatus*	31. *Epiglottis*
13. *Dorsal nasal meatus*	32. *Thyroid cartilage*
14. *Ventral nasal meatus*	33. *Vocal fold*
15. *Dorsal lateral nasal cartilage*	34. *Sternohyoideus*
16. *Alar fold*	35. *Cricoid cartilage*
17. *Nasolacrimal duct orifice*	36. *Laryngopharynx*
18. *Lyssa*	37. *Esophagus*
19. *Hard palate*	38. *Longus colli*

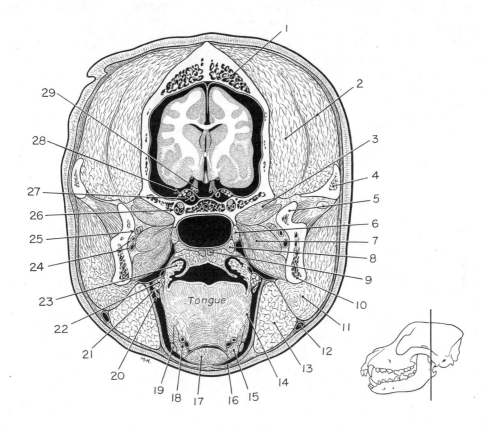

Figure 173. *Cross-section of head through palatine tonsil, looking rostrally.*

1. Diploe
2. Temporal m.
3. Lateral pterygoid m.
4. Zygomatic process of temporal bone
5. Condylar process
6. Tensor veli palatini
7. Medial pterygoid m.
8. Pterygopharyngeus m.
9. Palatinus m.
10. Mandible
11. Masseter m.
12. Facial vein
13. Digastricus m.
14. Styloglossus m.
15. Hyoglossus m.
16. Mylohyoideus m.

17. Geniohyoideus m.
18. Lingual a. and v.
19. Hypoglossal n.
20. Mandibular duct
21. Major sublingual duct
22. Sublingual salivary gland
23. Palatine tonsil in tonsillar fossa
24. Mandibular alveolar a. and v.
25. Mylohyoid, mandibular alveolar and lingual nn.
26. Maxillary a., v. and n. in alar canal
27. Internal carotid a. in cavernous sinus
28. Cranial nerves 3, 4 and 6, and ophthalmic division of 5
29. Cerebral arterial circle — caudal communicating a.

epiglottis. The **palatoglossal arch** is a fold on each side from the root of the tongue to the soft palate. The dorsal and ventral boundaries of the oropharynx are the soft palate and the root of the tongue. The lateral wall of the oropharynx contains the palatine tonsil in the tonsillar fossa.

The **palatine tonsil** is elongate and is located caudal to the palatoglossal arch. The medial wall of the fossa, which partially covers the tonsil, is the **semilunar fold.** The tonsil is attached laterally throughout its entire length. Reflect the tonsil from the fossa.

The nasal cavity extends from the nostrils to the choanae. It is divided into right and left halves by the nasal septum. Each half is divided into four meatuses. These were described with the bones of the nasal cavity and should be reviewed now. On the floor of the rostrolateral end of the ventral meatus, find the opening of the nasolacrimal duct. This duct comes from the lacrimal sac at the medial commissure of the eye.

The **nasopharynx** extends from the choanae to the junction of the palatopharyngeal arches. A **palatopharyngeal arch** extends caudally on each side from the caudal border of the soft palate to the dorsolateral wall of the nasopharynx. On the lateral wall of the nasopharynx, dorsal to the middle of the soft palate, is an oblique, slit-like opening, the pharyngeal opening of the auditory tube.

The **laryngopharynx** is dorsal to the larynx. It extends from the palatopharyngeal arches to the beginning of the esophagus. The esophagus begins at an annular constriction at the level of the cricoid cartilage, the **pharyngoesophageal limen.**

LARYNX

Cartilages. Study the cartilages of the larynx (Fig. 174) on specimens from which the muscles have been removed. Visualize their topography in your bisected specimen. The laryngeal cartilages include the paired arytenoid and cuneiform cartilages and the unpaired epiglottic, thyroid, cricoid, sesamoid and interarytenoid cartilages.

The **epiglottic cartilage** lies at the entrance to the larynx. Its lingual surface is attached to the basihyoid bone and faces the oropharynx. The apex lies just dorsal to the edge of the soft palate. The lateral margins are attached by mucosa to the cuneiform cartilage to form the aryepiglottic fold. Caudally the epiglottis attaches to the body of the thyroid cartilage.

The **thyroid cartilage** forms a deep trough which is open dorsally. The **rostral cornu** articulates with the thyrohyoid bone; the **caudal cornu** articulates with the cricoid cartilage. Ventrally the caudal border is notched by a median **caudal thyroid incisure.** The **cricothyroid ligament** attaches the caudal border to the ventral arch of the cricoid cartilage.

The **cricoid cartilage** forms a complete ring that lies partially within the trough of the thyroid cartilage. It has a wide dorsal plate, or lamina, and a narrow ventral arch. Near the caudal border at the junction of the lamina and the arch there is a lateral facet for articulation with the caudal cornu of the thyroid cartilage. On the cranial border of the lamina,

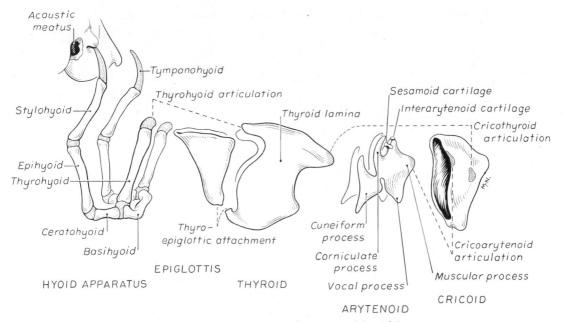

Figure 174. *Cartilages of larynx and hyoid bones.*

there is a prominent pair of lateral facets for articulation with the arytenoid cartilages.

The **arytenoid cartilage** is paired and is irregular in shape. Each articulates with a facet on the rostral border of the cricoid lamina and has a lateral **muscular process** and a ventrally directed **vocal process.** The **vocal fold** is attached between the vocal process and the thyroid cartilage. The arytenoid cartilage bears a **corniculate process** dorsally. Rostral to this process the **cuneiform process** is attached to the arytenoid (Fig. 174). The **vestibular fold** is attached to the ventral portion of the cuneiform process and forms the rostral boundary of the laryngeal ventricle.

The **glottis** consists of the vocal folds, the vocal processes of the arytenoid cartilages and the **rima glottidis,** which is the narrow passageway through the glottis. The **laryngeal ventricle** is a diverticulum of the larynx bounded laterally by the thyroid cartilage and medially by the

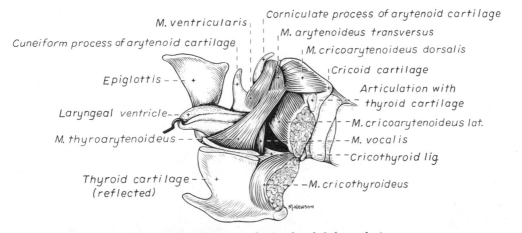

Figure 175. *Laryngeal muscles, left lateral view.*

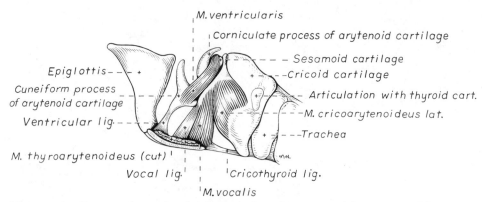

Figure 176.   *Laryngeal muscles, left lateral view, thyroarytenoideus, arytenoideus, transversus and cricoarytenoideus dorsalis removed.*

arytenoid cartilage. It opens into the larynx between the vestibular fold rostrally and vocal fold caudally.

Muscles.   The **cricothyroid muscle** (Figs. 175–177) lies ventral to the insertion of the sternothyroideus muscle and passes from the cricoid cartilage to the thyroid lamina. It tenses the vocal fold indirectly by drawing the ventral parts of the cricoid and thyroid cartilages together, thus abducting the arytenoid.

On the medial side of the specimen reflect the mucosa of the laryngopharynx.

The **cricoarytenoideus dorsalis** arises from the dorsolateral surface of the cricoid cartilage and inserts on the lateral surface of the arytenoid cartilage. Its function is to open the glottis. Transect the muscle and examine the articular surface of the cricoarytenoid joint.

The **cricoarytenoideus lateralis** arises from the lateral surface of the cricoid cartilage and inserts on the arytenoid cartilage between the cricoarytenoideus dorsalis and the vocalis. It acts to close the glottis.

The **thyroarytenoideus** is the parent muscle which gives rise to the vocalis and ventricularis muscles. It arises along the internal midline of the thyroid cartilage and inserts on the arytenoid cartilage. Its function is to relax the vocal fold and to constrict the glottis.

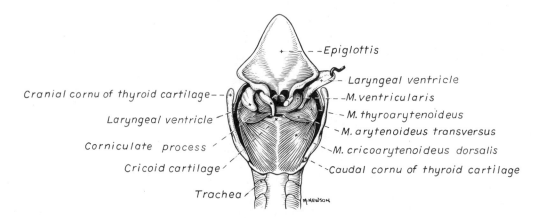

Figure 177.   *Laryngeal muscles, dorsal view.*

The **vocalis** is a medial division of the thyroarytenoid muscle. It arises on the internal midline of the thyroid cartilage and inserts on the vocal process of the arytenoid cartilage. Cut the laryngeal mucosa of the vocal fold and observe the medial side of this muscle. Attached along its rostral border is the **vocal ligament.**

The ventricularis and arytenoideus transversus will not be dissected.

Notice the relationship of the laryngeal ventricle to the laryngeal muscles. From its laryngeal opening it runs rostrally between the thyroid lamina and thyroarytenoideus laterally and the vestibular fold, the cuneiform process and the ventricularis medially.

THE EXTERNAL EAR

The external ear (Fig. 178) consists of the auricle and the external ear canal. The external ear canal is mostly cartilaginous but has a short osseous part. It extends to the tympanic membrane. It is not a straight canal but makes a right angle turn in the deeper part of its course.

Remove the skin from the base of the auricular cartilage. The **auricular cartilage** is funnel-shaped. Its external, convex surface faces caudally; the internal, concave surface faces rostrally. It has a slightly folded medial and lateral margin called the **helix.** The auricular cartilage is thin and pliable except proximally, where it thickens and rolls into a tube. On

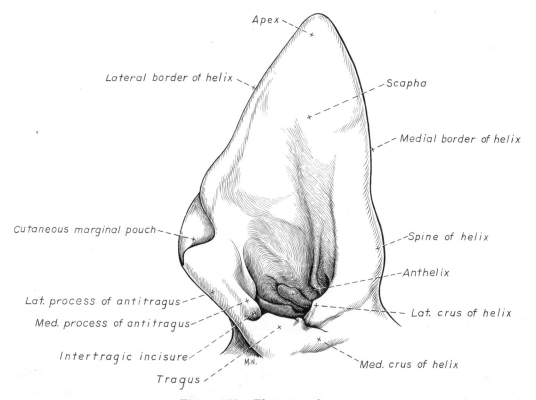

Figure 178.   The external ear.

its internal, concave wall at the level of the beginning of the ear canal there is a transverse ridge, the **anthelix.** The entire concave, triangular area between the helix and the anthelix is the **scapha.** Opposite the anthelix, the rostral boundary of the initial part of the ear canal is formed by a thick, quadrangular plate of cartilage, the **tragus.** Projecting caudally from the tragus and completing the lateral boundary of the external ear canal is a thin, elongate piece of cartilage, the **antitragus.** The **intertragic incisure** separates these two parts of the auricular cartilage.

The lateral portion of the helix is indented proximally by an incisure. At this point the skin forms a pouch, the **marginal cutaneous sac.**

The medial helix is nearly straight. An abrupt angle in this border proximally forms the **spine of the helix.** Between the spine of the helix and the tragus the medial border of the ear canal is formed by two curved portions of the cartilage, the **medial** and **lateral crura** of the helix. They both end laterally in a free border separated from the tragus by the **tragohelicine incisure.**

Incise the lateral wall of the ear canal to observe its course to the tympanic membrane.

Interposed between the auricular cartilage and the external acoustic meatus of the temporal bone is the **annular cartilage.** This is a band of cartilage which overlaps the bony projection of the meatus.

MUSCLES OF MASTICATION AND RELATED MUSCLES

Cut all attachments of the temporal and masseter muscles to the zygomatic arch, and remove it.

The **temporalis muscle** (Fig. 173) arises from the temporal fossa and inserts on the coronoid process of the mandible. Remove the muscle from its origin by scraping it off the bone with a blunt instrument such as the scalpel handle and reflect it. The temporal and masseter muscles fuse between the zygomatic arch and the coronoid process.

The **masseter muscle** arises from the zygomatic arch, where its deep portion is intermingled with the fibers of the temporal muscle. It inserts in the masseteric fossa and the ventrolateral surface of the ramus of the mandible. The muscle is covered by a strong, glistening aponeurosis and contains many tendinous intermuscular strands.

Transect the temporal muscle as close to its insertion as possible. With bone cutters remove the coronoid process and the attached muscles. Observe the dorsal surface of the pterygoid muscles, which are now exposed under the orbit.

Between the eyeball and the pterygoid muscles is the **zygomatic salivary gland.** The gland opens into the vestibule by one main duct lateral to the last upper molar tooth and two to four minor ducts caudal to the main one.

The **medial** and **lateral pterygoid muscles** arise from the pterygopalatine fossa and insert on the medial surface and caudal margin of the mandible ventral to the insertion of the temporal muscle. On the medial side of your specimen, cut the mucosa of the oropharynx from the rostral end of the palatine tonsil to the midline at the junction of the soft and hard palates. Reflect the cut edges to expose the ventral surface of the medial pterygoid muscle.

The function of the temporal, masseter and pterygoid muscles is to close the jaws.

Reflect the mandibular and parotid salivary glands to expose the digastricus muscle.

The **digastricus** arises from the jugular process of the occipital bone and is inserted on the angle and body of the mandible. A tendinous intersection crosses its belly and divides it into rostral and caudal parts. Transect it and expose its attachments. It acts to open the jaws.

LINGUAL MUSCLES

The muscles of the tongue (Figs. 172, 179) may be divided into extrinsic and intrinsic groups. Three paired extrinsic muscles enter the tongue. The styloglossus and hyoglossus muscles are best exposed on the lateral side, the genioglossus on the medial side.

The **styloglossus** arises from the stylohyoid bone, passes rostroventrally lateral to the palatine tonsil and inserts in the middle of the tongue. It retracts and elevates the tongue.

The **hyoglossus** arises from the thyrohyoid and the basishyoid and passes into the root of the tongue. It lies medial to the styloglossus and retracts and depresses the tongue.

The **genioglossus** arises from the symphysis and adjacent surface of the body of the mandible. It joins its fellow at the median plane and is

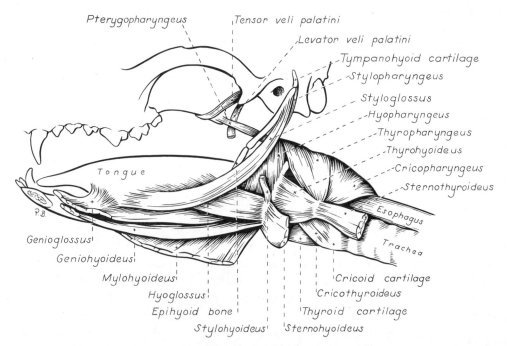

Figure 179. Muscles of pharynx and tongue.

bounded laterally by the geniohyoideus and the hyoglossus. Its caudal fibers protrude the tongue and its rostral ones retract the apex. It lies partly in the frenulum.

HYOID MUSCLES

The hyoid muscles (Fig. 179) are associated with the hyoid apparatus, which suspends the larynx and anchors the tongue. They function in swallowing, lolling, lapping and retching. All muscles of this group have names with the suffix -*hyoideus*. The prefixes of the hyoid muscles designate the bone or part from which they arise. Dissect the following hyoid muscles from the lateral side.

The **sternohyoideus,** from its origin on the sternum and first costal cartilage, is fused to the deeper sternothyroideus for the first third of its length. It then separates from this muscle and runs an independent course to insert on the basihyoid bone. Its origin was previously dissected.

The **thyrohyoideus** is a short muscle which lies dorsal to the sternohyoideus. It extends from the thyroid cartilage of the larynx to the thyrohyoid bone.

The **mylohyoideus** spans the intermandibular space. It arises as a thin sheet of transverse fibers from the medial surface of the body of the mandible. It is inserted on its fellow muscle at the midventral raphe. Caudally it inserts on the basihyoid. It forms a sling which aids in the support of the tongue.

The **geniohyoideus** lies deep to the mylohyoideus. It is a muscular strap which arises on and adjacent to the mandibular symphysis. It parallels its fellow along the medial plane and attaches to the basihyoid. Contraction of the geniohyoideus draws the hyoid apparatus and larynx rostrally.

PHARYNGEAL MUSCLES

The pharyngeal muscles (Figs. 172, 179) aid directly in swallowing.

The **cricopharyngeus** arises from the lateral surface of the cricoid cartilage. Its fibers are inserted on the median dorsal raphe of the laryngopharynx. Caudally its fibers blend with the esophagus.

The **thyropharyngeus** arises from the lateral side of the thyroid lamina and is inserted on the median dorsal raphe of the pharynx. This muscle is rostral to the cricopharyngeus and caudal to the hyopharyngeus.

The **hyopharyngeus** is in two parts as it arises from the lateral surface of the thyrohyoid bone and the keratohyoid bone. This origin was previously transected. The fibers of both parts form a muscle plate which passes upward over the larynx and pharynx to be inserted on the median dorsal raphe of the pharynx with its fellow from the opposite side.

The remaining pharyngeal muscles and muscles of the palate need not be dissected. (See Fig. 172.)

The **palatopharyngeus** passes from the soft palate into the lateral and dorsal wall of the pharynx. The **pterygopharyngeus** arises from the pterygoid bone, passes caudally, and is inserted in the dorsal wall of the pharynx. These muscles constrict and shorten the pharynx.

The **stylopharyngeus** arises from the stylohyoid bone and passes caudolaterally deep to the hyopharyngeus and thyropharyngeus muscles to be inserted in the dorsolateral wall of the pharynx. It acts to dilate the pharynx.

The **levator veli palatini** arises from the tympanic part of the temporal bone and passes ventrally to enter the soft palate caudal to the pterygoid bone. It raises the caudal end of the soft palate (Fig. 172).

The **tensor veli palatini** arises mainly from the cartilaginous wall of the auditory tube and is inserted on the pterygoid bone and, medially, in the soft palate.

JOINTS OF THE HEAD

The **temporomandibular joint** is the articulation between the condyloid process of the mandible and the mandibular fossa of the temporal bone. A thin articular meniscus lies between these two bony surfaces. The meniscus is completely attached peripherally to the joint capsule, forming two separate compartments in the joint. Study this joint on the left side where the zygomatic arch was removed.

The **atlanto-occipital joint** was disarticulated when the head was removed and bisected. It is formed by the occipital condyles and the corresponding foveae of the atlas. Dorsally the space between the dorsal edge of the foramen magnum and the arch of the atlas is covered by the dorsal atlanto-occipital membrane.

THE EYE AND RELATED STRUCTURES

The **periorbita** is a fibrous, cone-shaped membrane that encloses the eyeball and its muscles, vessels and nerves. Where the periorbita contacts the bone medially it is the periosteum of the orbit. Its apex is caudal and it attaches to the bony margin of the optic canal and the orbital fissure. Rostrally it widens to blend with the periosteum of the face. Orbital fat may be observed on both sides of the periorbita.

Reflect the orbital ligament and the periorbita from the dorsolateral surface of the eyeball. The **lacrimal gland** lies ventral to the zygomatic process of the frontal bone. Small ducts that cannot be seen grossly empty

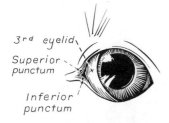

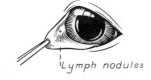

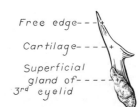

Figure 180.   Third eyelid.

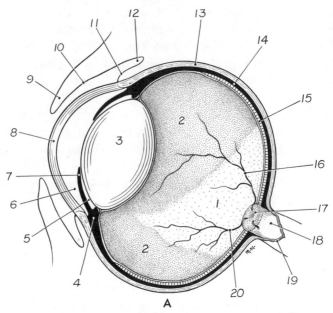

Figure 181a.   *Sagittal section of the eyeball.*

1. *Tapetum lucidum*
2. *Tapetum nigrum*
3. *Lens*
4. *Ciliary body*
5. *Posterior chamber*
6. *Anterior chamber*
7. *Iris*
8. *Cornea*
9. *Upper eyelid*
10. *Palpebral conjunctivum*
11. *Bulbar conjunctivum*
12. *Fornix*
13. *Sclera*
14. *Choroid*
15. *Retina*
16. *Superior retinal vein*
17. *Optic disc*
18. *Optic nerve*
19. *Dura*
20. *Inferior medial retinal vein*

their secretion into the conjunctival sac at the dorsal fornix. Incise the periorbita longitudinally and reflect it to expose the eyeball muscles and the levator of the upper eyelid (Fig. 182).

The **levator palpebrae superioris** lies superficial to the extrinsic muscles. It begins at the apex of the orbit, extends medial to the dorsal rectus and widens to insert as a flat tendon in the upper eyelid.

There are seven extrinsic muscles of the eyeball: two **obliquus muscles,** four **rectus muscles** and the **retractor bulbi.** All of these extrinsic muscles insert on the fibrous coat of the eyeball, the sclera, near the equator of the eyeball. The rectus muscles insert closer to the corneoscleral junction than the retractor muscles.

The four rectus muscles are the **dorsal rectus, medial rectus, ventral rectus** and **lateral rectus.** As they course rostrally from their small area of origin around the optic canal, they diverge and attach to the sclera on an imaginary line encircling the eyeball at its equator. In the spaces between the rectus muscles parts of the retractor bulbi can be seen. The retractor bulbi muscle consists of four fascicles that surround the optic nerve, a dorsal pair and a ventral pair.

The **dorsal oblique** ascends on the dorsomedial side of the extraocular muscles dorsal to the medial rectus. It is a narrow muscle that forms a

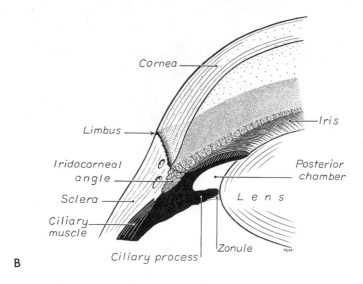

Figure 181b. *Cross-section of the eyeball at the iridocorneal angle.*

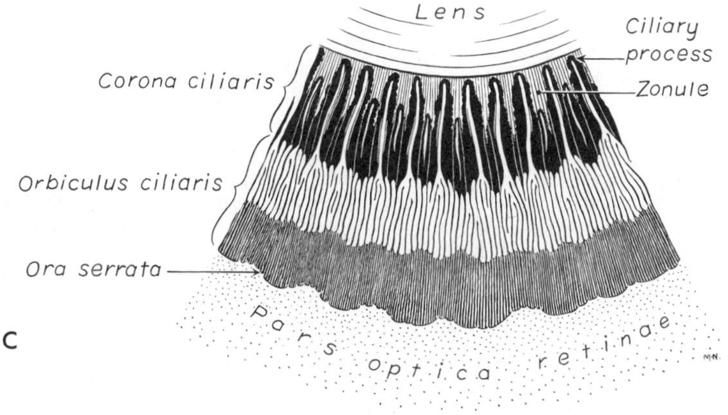

Figure 181c. *Inner surface of a segment of the ciliary body and zonule.*

long tendon rostrally that passes through a groove in the trochlea. The **trochlea** is a cartilaginous plaque attached at the level of the medial angle of the eye to the wall of the orbit. The tendon of the dorsal oblique muscle turns and courses laterally after passing around the trochlea and attaches to the sclera under the tendon of insertion of the dorsal rectus muscle. Transect and reflect the dorsal rectus and levator palpebrae superioris to expose the insertion of the dorsal oblique.

The **ventral oblique** arises from the rostral border of the palatine bone at the level of the maxillary foramen, passes ventral to the ventral rectus and is inserted on the sclera at the insertion of the lateral rectus.

In order to understand the action of these individual muscles, consider the eyeball as having three imaginary axes that cross in the center of the globe. The dorsal and ventral rectus muscles would rotate the eyeball

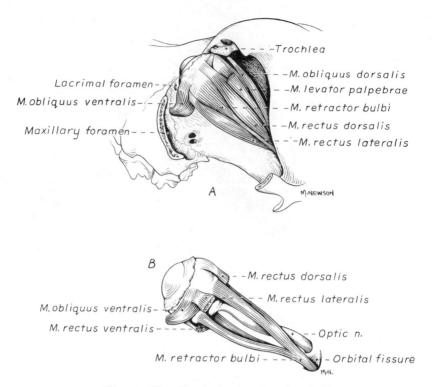

Figure 182. Extrinsic muscles of eyeball.

around a horizontal axis through the equator from medial to lateral. The medial and lateral rectus muscles would rotate it around a vertical axis through the equator from dorsal to ventral. The oblique muscles would rotate the eyeball around a longitudinal axis passing from rostral to caudal through the center of the eyeball.

Enlarge the opening (rima) between the upper and lower eyelids by cutting through the lateral commissure and the underlying conjunctiva to the eyeball. This will facilitate exposure of the third eyelid in the medial angle. Elevate the third eyelid and observe the lymph nodules on its bulbar conjunctival surface and the superficial gland of the third eyelid.

BULBUS OCULI

In the following dissection leave the bulbus oculi, or eyeball, in the orbit. Directional terms used with reference to the eyeball are anterior and posterior, and superior and inferior. The wall is composed of three layers (Fig. 181).

1. The **external fibrous coat** is composed of the **cornea,** forming the anterior one-sixth, and the **sclera,** forming the posterior five-sixths. The cornea is transparent and circular. It meets the dense, opaque sclera peripherally at the corneoscleral junction or **limbus** of the cornea. The sclera is dull gray-white in color. Anteriorly it is covered by bulbar conjunctiva. Posterior to this the extrinsic ocular muscles insert in its wall, and it is penetrated by blood vessels and by nerves including the

large optic nerve posteriorly. Clean the muscles from the surface of the eyeball. Note the point of penetration by the optic nerve.

2. The **middle vascular coat** (the **uvea**) consists of three continuous parts from posterior to anterior: the choroid, ciliary body and iris. The **iris** can be seen through the cornea as a brownish diaphragm with a central opening, the **pupil.** With a sharp scalpel or razor blade make a sagittal cut through the eyeball at the limbus.

The **choroid** is the posterior portion of the vascular coat which is pigmented and which lines the internal surface of the sclera as far anterior as the level of the lens. The light-colored, reflective triangular area in the caudodorsal part of the fundus is the **tapetum lucidum** of the choroid. The choroid is firmly attached to the sclera around the optic nerve.

The vascular coat forms a thick circular mound at the level of the limbus. This is the **ciliary body.** The ciliary body contains numerous muscle bundles that function in the regulation of the shape of the lens. The internal surface of the ciliary body is marked by longitudinal folds, the **ciliary processes.** Observe these on the lateral portion of the eyeball that was removed.

The **zonule** is a gel that contains numerous fine fibers, **zonular fibers,** that pass from the ciliary processes to the equator of the lens. Contraction of the ciliary muscles relaxes the tension of the zonule on the lens, allowing the elastic lens to become more spherical in accommodation.

The vascular coat is continued anterior to the ciliary body by the iris. The iris regulates the size of the pupil.

The **lens** in a fixed specimen is firm and opaque. In life it is transparent and elastic. It is bounded posteriorly by the transparent jelly-like **vitreous body,** which fills the **vitreous chamber** posterior to the lens. Anteriorly it is bounded by the iris and by aqueous humor. The **aqueous humor** fills the space between the cornea and the lens. This space is divided into two chambers. The **anterior chamber** is the space between the cornea and the iris. The **posterior chamber** is a narrow cavity between the iris and the lens. Remove the lens with a forceps. Note the zonular fibers as they stretch and break.

3. The internal or **nervous coat** of the eye is the **retina.** The portion of the retina containing the elements that are sensitive to light stimulation is the **pars optica.** It lines the internal surface of the choroid to the level of the ciliary processes. From this boundary anteriorly, a thin non-nervous portion of the retina covers the internal surface of the ciliary body, the **pars ciliaris,** and the posterior aspect of the iris, the **pars iridica.** The boundary between the nervous and non-nervous portions of the retina is called the **ora serrata.** Except for the area of the **tapetum lucidum,** the pars optica retina appears black, **tapetum nigrum,** as a result of pigment in the retina and choroid.

Notice the entrance of the optic nerve into the posterior aspect of the eyeball. This is the **optic disc.** With careful observation of the disc you may see the retinal vessels that enter with it to supply the internal surface of the retina.

SUPERFICIAL VEINS OF THE HEAD

The **external jugular vein** is formed by the confluence of the linguo-facial and maxillary veins caudal to the mandibular salivary gland, which lies between these two veins.

The **lingual vein** is the first large tributary that enters the **linguofacial vein** ventrally. Its radicles drain blood from the tongue, the larynx and part of the pharynx. These radicles will not be dissected.

The **facial vein** is the other tributary of the linguofacial. The radicles which form the facial vein lie on the dorsal surface of the muzzle. One of these, the **dorsal nasal,** runs caudally from the nose, while another, the **angularis oculi,** passes rostrally from the orbit. Blood may drain from the face in either direction through the angularis oculi. The remaining branches will not be dissected for to do so would be to sacrifice nerves and arteries. These branches drain the lateral muzzle and the upper and lower lips.

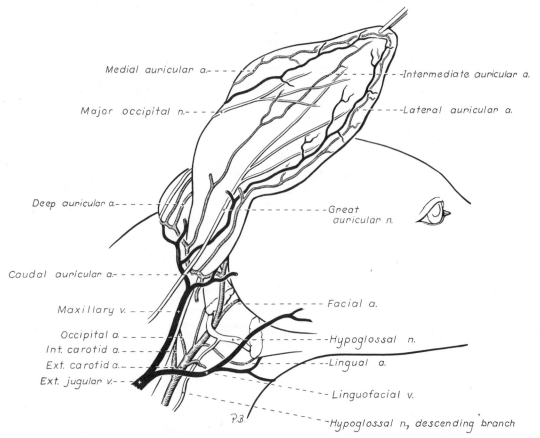

Figure 183. *Vessels and nerves of external ear.*

The **maxillary vein** drains the ear, orbit, palate, cheek and mandible as well as the cranial cavity. It will not be dissected.

FACIAL NERVE

The **facial** or **seventh cranial nerve** (Fig. 184) supplies all of the superficial muscles of the head. The nerve courses through the facial canal of the petrous temporal bone, leaves the skull at the stylomastoid foramen just caudal to the external acoustic meatus and divides into several branches.

Reflect the parotid gland. Dissect between the parotid and sublingual glands to expose the facial nerve arising from the stylomastoid foramen caudal to the ear canal. The maxillary vein may be transected as it crosses the lateral surface of the nerve at this site. Dissect the following branches of the facial nerve.

The **auriculopalpebral nerve** arises as the facial nerve curves rostrally ventral to the base of the ear. Dissect deep to the rostral edge of the parotid gland to locate the nerve. **Rostral auricular branches** course through the parotid gland and are distributed to the rostral auricular muscles. The auriculopalpebral nerve crosses the zygomatic arch and

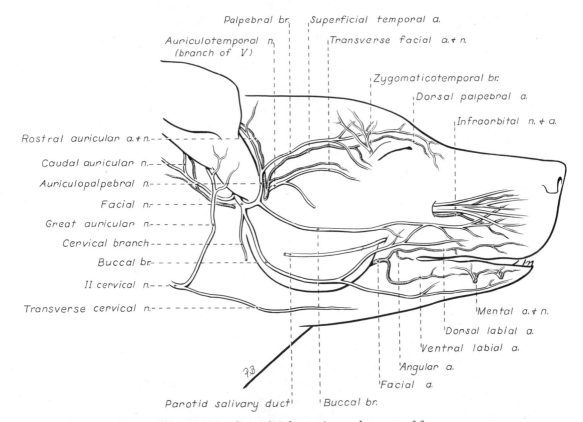

Figure 184. *Superficial arteries and nerves of face.*

supplies branches to the rostral auricular plexus and continues to the orbit to supply **palpebral branches** to the orbicularis oculi. Beyond this a branch passes medial to the orbit and continues rostrally on the nose to supply the muscles of the nose and upper lip.

Two **buccal branches** course across the masseter muscle to innervate the muscles of the cheek, the upper and lower lips and the lateral surface of the nose.

The **auriculotemporal nerve,** a branch of the mandibular nerve from the trigeminal or fifth cranial nerve, is apparent in the dissection of the facial nerve. It emerges between the caudal border of the masseter muscle and the base of the external ear below the zygomatic arch. It may be found deep to the origin of the auriculopalpebral nerve. The auriculotemporal nerve supplies sensory branches to the skin of the external ear and the temporal, zygomatic and masseteric regions.

CERVICAL STRUCTURES

The **thyroid gland** is dark-colored and consists of two separate lobes lying lateral to the first five tracheal rings.

There are two **parathyroid glands** associated with each thyroid lobe. They are light-colored, spherical bodies. The **external parathyroid** most commonly lies in the fascia at the cranial pole of the thyroid lobe. It may be entirely separate from the thyroid tissue or embedded in the cranial pole of the thyroid, external to its capsule. The **internal parathyroid** lies deep to the thyroid capsule on the medial aspect of the lobe. Occasionally it is embedded in the parenchyma of the thyroid. The location of these glands is subject to variation.

The **cervical portion of the esophagus** extends from the pharynx to the thoracic portion of the esophagus at the thoracic inlet. It begins opposite the middle of the axis dorsally and the caudal border of the cricoid cartilage ventrally. A plicated ridge of mucosa, the **pharyngoesophageal limen,** marks the boundary between the laryngopharynx and the esophagus. The esophagus inclines to the left so that at the thoracic inlet it lies to the left of the trachea.

The **trachea** extends from a transverse plane through the middle of the axis to a plane between the fourth and fifth thoracic vertebrae. It is composed of approximately 35 C-shaped **tracheal cartilages.** They are open dorsally and the space is bridged by the tracheal muscle.

COMMON CAROTID ARTERY

On the right side expose the **common carotid artery** in the carotid sheath and observe the following branches.

The **cranial thyroid artery** (Fig. 185) arises from the ventral surface of the common carotid at the level of the larynx. It supplies the thyroid and parathyroid glands, the pharyngeal muscles, the laryngeal muscles

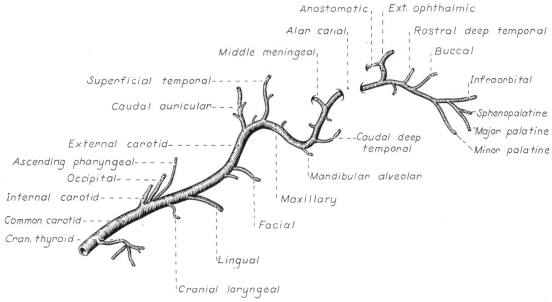

Figure 185. *Branches of common carotid artery.*

and mucosa, the cervical parts of the trachea and esophagus and the adjacent portions of the sternocephalicus and cleidomastoideus. Clean the origin of this vessel.

The large **medial retropharyngeal lymph node** is dorsal to the common carotid artery and ventral to the wing of the atlas. Afferent vessels arise from the tongue, the nasal cavity, the pharynx, the salivary glands, the external ear, the larynx and the esophagus. The tracheal duct of each side arises from this lymph node. Identify the internal and external carotid arteries, the terminal branches of the common carotid artery.

The **internal carotid artery** is closely associated with the occipital artery. A bulbous enlargement at the origin of the internal carotid artery is the **carotid sinus,** a baroreceptor. (The carotid body, a chemoreceptor, lies at the bifurcation of the carotid arteries.) Beyond this the vessel ascends across the lateral surface of the pharynx medial to the occipital artery. No branches leave the internal carotid in its extracranial course. It enters the carotid canal in the depths of the occipitotympanic fissure. Its course from here into the cranial cavity has been described. Its branches to the brain will be dissected later.

The **external carotid artery** (Fig. 185) passes cranially medial to the digastricus. At the caudal border of the mandible, rostroventral to the annular cartilage of the ear, the vessel terminates by dividing into the **superficial temporal** and **maxillary** arteries. The maxillary is the direct continuation of the external carotid. Dissect the following branches of the external carotid.

The **lingual artery** leaves the ventral surface of the external carotid and passes rostrally to supply the tonsil and tongue.

The **facial artery** leaves the external carotid beyond the lingual. A branch, the **sublingual artery,** lies medial to the digastric muscle and is accompanied by the mylohyoid nerve. It runs rostrally into the tongue.

The facial artery courses rostrally between the digastric and masseter muscles to reach the cheek lateral to the mandible, where it supplies the lips and nose.

The **caudal auricular artery** arises from the external carotid at the base of the ear and ascends under the caudal auricular muscles. Reflect the caudal limb of the parotid gland to expose the caudal auricular artery and its branches, which need not be dissected. A lateral, intermediate and medial auricular artery course distally on the convex surface of the ear.

The **superficial temporal artery** arises rostral to the base of the auricular cartilage at the caudodorsal border of the mandible and courses dorsally. It supplies the parotid gland, the masseter and temporal muscles, the rostral auricular muscles and the eyelids.

The **maxillary artery** is the larger terminal branch of the external carotid artery. It is deeply placed and is closely associated with a number of cranial nerves.

Remove the auricular muscles and reflect the ear caudally. Cut through the origin of the temporal muscle along its margin. With a blunt instrument remove it from the temporal fossa. Cut the attachments of the temporal and masseter muscles to both sides of the zygomatic arch. Sever the orbital ligament at the arch. With bone cutters detach each end of the arch and remove it. Transect the temporal muscle as close to its insertion on the coronoid process as possible. Cut off the coronoid process below the level of the ventral border of the zygomatic arch with bone cutters. Remove the temporal muscle to expose the periorbital tissues and the pterygoid muscles. Vessels and nerves entering the temporal muscle must be severed. Loosen the temporomandibular joint by forcing the symphyseal end of the mandible medially. Rotate the mandible so that the stump of the coronoid process is forced laterally.

In the oral cavity reflect the mucosa from the level of the root of the tongue to the frenulum along the sublingual fold.

TRIGEMINAL NERVE

Mandibular Nerve. The branches of the mandibular nerve from the trigeminal or fifth cranial nerve have been exposed by this dissection (Fig. 186). The mandibular nerve leaves the cranial cavity via the oval foramen. Branches arise on the surface of the pterygoid muscles ventral and lateral to the apex of the periorbita. These include pterygoid, deep temporal, and masseteric nerves that innervate the muscles of mastication. The buccal nerve crosses the pterygoid muscles and is sensory to the mucosa and skin of the cheek.

Rotate the stump of the coronoid process laterally to observe the mandibular alveolar, mylohyoid and lingual nerves, which cross the dorsal surface of the medial pterygoid muscle.

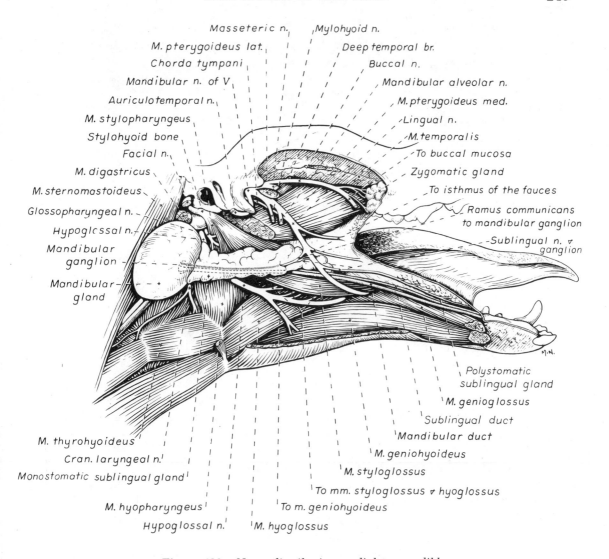

Masseteric n.
M. pterygoideus lat.
Chorda tympani
Mandibular n. of V
Auriculotemporal n.
M. stylopharyngeus
Stylohyoid bone
Facial n.
M. digastricus
M. sternomastoideus
Glossopharyngeal n.
Hypoglossal n.
Mandibular ganglion
Mandibular gland

Mylohyoid n.
Deep temporal br.
Buccal n.
Mandibular alveolar n.
M. pterygoideus med.
Lingual n.
M. temporalis
To buccal mucosa
Zygomatic gland
To isthmus of the fauces
Ramus communicans to mandibular ganglion
Sublingual n. & ganglion

Polystomatic sublingual gland
M. genioglossus
Sublingual duct
Mandibular duct
M. geniohyoideus
M. styloglossus
To mm. styloglossus & hyoglossus
To m. geniohyoideus
M. hyoglossus

M. thyrohyoideus
Cran. laryngeal n.
Monostomatic sublingual gland
M. hyopharyngeus
Hypoglossal n.

Figure 186. Nerve distribution medial to mandible.

The **mandibular alveolar nerve** (sensory) enters the mandibular foramen of the lower jaw. It courses through the mandibular canal supplying sensory nerves to the teeth. The mental nerves that supply the lower lip are branches from this nerve.

The **mylohyoid nerve** (motor) is caudal to the mandibular alveolar. It reaches the ventral border of the mandible, supplies a branch to the rostral belly of the digastricus muscle and continues into the mylohyoid muscle. Observe this nerve emerging on the medial side of the angle of the mandible lateral to the mylohyoideus. Trace its distribution to the digastric and mylohyoid muscles.

The **lingual nerve** (sensory) may be observed between the mylohyoid and the styloglossus. It crosses the lateral side of the mandibular and sublingual ducts and enters the tongue. It is sensory to the rostral two-thirds of the tongue.

The **auriculotemporal nerve** (sensory) leaves the mandibular nerve at the oval foramen, passes caudal to the retroarticular process of the temporal bone and emerges between the base of the auricular cartilage and the masseter muscle, where it was previously seen.

MAXILLARY ARTERY

Complete the disarticulation of the temporomandibular joint and remove the lateral pterygoid muscle. Reflect the ramus of the mandible laterally and identify the following branches of the maxillary artery (Fig. 185). The first three arise before the maxillary artery enters the alar canal.

1. The **mandibular alveolar artery** enters the mandibular foramen and courses through the mandibular canal. It supplies branches to the roots of the teeth in the lower jaw.

2. The **caudal deep temporal artery** arises near the mandibular alveolar artery and enters the temporal muscle. Only the origin of this artery may be seen.

3. The **middle meningeal artery** passes through the oval foramen, where it will be followed in a later dissection to the dura over the cerebral hemispheres. Do not dissect its origin.

4. The **external ophthalmic artery** arises from the maxillary upon its emergence from the alar canal and penetrates the apex of the periorbita. The external ophthalmic artery gives rise to the vessels that supply the structures within the periorbita. Incise the periorbita longitudinally along its dorsolateral border and reflect it.

The branches of the external ophthalmic artery need not be dissected. One anastomotic branch passes caudally through the orbital fissure to join the internal carotid and middle meningeal arteries within the cranial cavity. Another anastomotic branch joins the internal ophthalmic artery emerging from the optic canal on the optic nerve. From this anastomosis vessels are supplied to the eyeball. Branches of the external ophthalmic artery supply the extrinsic muscles of the eyeball and the lacrimal gland. The **external ethmoidal artery** passes dorsal to the extraocular muscles and enters an ethmoidal foramen. Within the cranial cavity it joins with the internal ethmoidal artery and supplies the cribriform plate, the ethmoid labyrinth and the nasal septum.

5. Among the terminal branches of the maxillary artery are

The **minor palatine artery,** which passes ventrally caudal to the hard palate and is distributed to the adjacent soft and hard palates. Clean the mucosa from the palate just medial to the last molar and see the branches of this vessel.

The **major palatine artery,** which enters the caudal palatine foramen and passes through the major palatine canal to supply the hard palate.

The **sphenopalatine artery,** which passes through the sphenopalatine foramen, to the interior of the nasal cavity.

The **infraorbital artery,** which supplies dental branches to the caudal cheek teeth, enters the maxillary foramen and passes through the infraorbital canal. Within the canal dental branches arise which supply the premolars and the canine and incisor teeth. The infraorbital artery terminates as the lateral and rostral dorsal nasal arteries, which supply the nose and the upper lip.

CRANIAL NERVES

There are twelve pairs of cranial nerves (Figs. 187–189). Each pair is both numbered and named. The numbers indicate the order in which they arise from the brainstem and the names are descriptive. Some of these have already been dissected, others will be dissected now or observed on demonstrations.

1. The **olfactory** or **first cranial nerve** consists of numerous nerves that arise in the olfactory epithelium of the caudal nasal mucosa and pass through the cribriform foramina to the olfactory bulbs. No dissection is necessary.

2. The **optic** or **second cranial nerve** is surrounded by the retractor

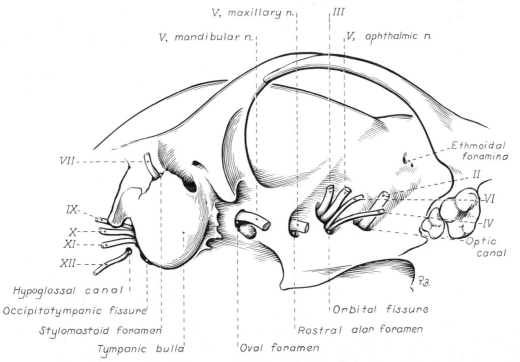

Figure 187. *Cranial nerves leaving skull, ventrolateral view.*

bulbi muscle within the periorbita. Observe the nerve as it enters the optic canal.

3. The **oculomotor** or **third cranial nerve** passes through the orbital fissure and enters the periorbita with the optic nerve. It innervates the dorsal, medial and ventral rectus, the ventral oblique and the levator palpebrae superioris. The **ciliary ganglion** is an irregular enlargement at the termination of the oculomotor nerve. It contains parasympathetic cell bodies of postganglionic axons that innervate the sphincter pupillae of the iris. Study this nerve on the demonstration specimen.

4. The **trochlear** or **fourth cranial nerve** enters the periorbita through the orbital fissure. It innervates the dorsal oblique. Observe it on the demonstration specimen.

5. The **trigeminal** or **fifth cranial nerve** divides into three nerves as it emerges from the trigeminal canal in the petrosal bone: the ophthalmic, the maxillary and the mandibular. The mandibular nerve has been dissected.

The **ophthalmic nerve** passes through the orbital fissure and supplies sensory nerves that enter the periorbita. These need not be dissected. The **frontal** and **infratrochlear nerves** pass rostrally between the dorsal oblique and dorsal rectus muscles to innervate the upper eyelid. Long

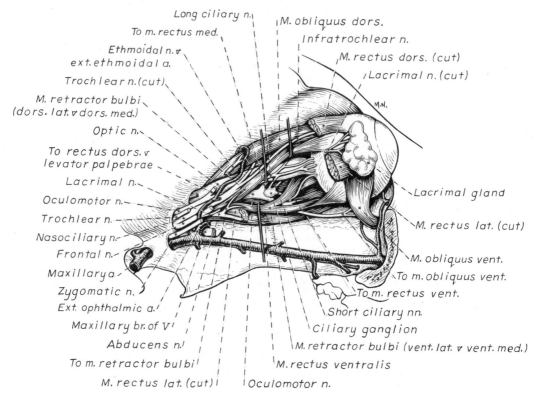

Figure 188. Nerves of eyeball and its adnexa.

ciliary nerves follow the optic nerve to the eyeball. The **ethmoidal nerve** passes through an ethmoidal foramen and the cribriform plate to inner- vate the nasal mucosa and skin of the nose.

The **maxillary nerve** enters the alar canal via the round foramen. It emerges from the rostral alar foramen and crosses the pterygopalatine fossa dorsal to the pterygoid muscles accompanied by the maxillary ar- tery. Dissect the following branches.

The **zygomatic nerve** enters the periorbita and divides into two branches that pass rostrally to innervate the lacrimal gland and the lateral portion of the upper and lower eyelids.

The **pterygopalatine ganglion** is dorsal to the maxillary nerve on the surface of the middle pterygoid muscle. It contains cell bodies of postganglionic parasympathetic axons that supply the lacrimal, nasal, and palatine glands. The postganglionic axons course with the branches of the maxillary nerve to their terminations.

The **pterygopalatine nerve** arises beyond the level of the ptery- gopalatine ganglion and divides into three nerves: the **minor** and **major palatine nerves** of the palate and the **caudal nasal nerve** of the nasal mucosa. Dissect their origin.

The **infraorbital nerve** is the continuation of the maxillary nerve in the pterygopalatine fossa. It enters the infraorbital canal via the maxillary foramen. Along its course through the infraorbital canal it gives off **maxillary alveolar branches** which supply the roots of the teeth. As the infraorbital nerve emerges from the infraorbital foramen it divides into a number of fasciculi which are distributed to the skin and adjacent structures of the upper lip and nose. Dissect these branches as they emerge from the infraorbital foramen.

6. The **abducens** or **sixth cranial nerve** passes through the orbital fissure and enters the periorbita. It innervates the retractor bulbi and the lateral rectus. Observe this nerve on the demonstration specimen.

7. The **facial** or **seventh cranial nerve** enters the internal acoustic meatus of the petrosal bone. It courses through the facial canal and emerges through the stylomastoid foramen. Its sensory ganglion, the **geniculate ganglion,** is located within the facial canal. Within the canal, the facial nerve gives off small branches that pass through the middle ear and enter the pterygopalatine fossa. These are preganglionic parasympa- thetic axons that are motor to the mandibular and sublingual salivary gland and the lacrimal, nasal and palatine glands. Those neurons to the lacrimal, nasal and palatine glands synapse in the pterygopalatine gan- glion and are distributed to these glands via branches of the maxillary nerve. The neurons to the salivary glands are in the **chorda tympanic nerve.** This nerve emerges from the rostral side of the tympanic bulla and joins the lingual nerve, by which it is distributed to the mandibular and sublingual ganglia. These ganglia are closely associated with the glands that are innervated. No dissection is required. The peripheral branches of the facial nerve have been dissected.

8. The **vestibulocochlear** or **eighth cranial nerve** enters the internal acoustic meatus with the facial nerve. It terminates in the membranous labyrinth of the inner ear. It is the sensory nerve concerned with balance and hearing and will be dissected with the brain.

9. The **glossopharyngeal** or **ninth cranial nerve** passes through the jugular foramen and the occipitotympanic fissure. The distal ganglion of the glossopharyngeal is sensory and is located in the region of the jugular foramen. In its course through the fissure the glossopharyngeal gives off small branches that pass through the middle ear. Some of these reach the **otic ganglion,** located near the oval foramen. These are preganglionic neurons that synapse on parasympathetic cell bodies of postganglionic axons in the otic ganglion. By means of adjacent branches of the trigeminal they are distributed to the parotid and zygomatic salivary glands. No dissection is necessary. Beyond the fissure the glossopharyngeal crosses the lateral surface of the cranial cervical ganglion and divides into pharyngeal and lingual branches that are sensory to the pharyngeal mucosa and motor to the stylopharyngeus and other pharyngeal muscles. In addition some branches course to the carotid sinus and others contribute to the pharyngeal plexus along with branches of the vagus nerve.

10. The **vagus** or **tenth cranial nerve** passes through the jugular foramen and the tympanooccipital fissure. It courses along the common carotid artery in the neck with the sympathetic trunk and through the thorax on the esophagus to terminal branches in the thorax and abdomen (which have already been dissected). The following branches are distributed to cranial cervical structures.

There are two ganglia associated with this nerve. The **proximal ganglion** of the vagus lies in the jugular foramen. The **distal ganglion** of the vagus is found outside the occipitotympanic fissure, ventral and medial to the tympanic bulla. These are sensory ganglia. The distal ganglion contains the cell bodies of the visceral afferent neurons that are distributed to most of the viscera of the body. Beyond the distal ganglion the vagus joins the sympathetic trunk, with which it remains associated throughout its cervical course to the thoracic inlet. Branches from the vagus and glossopharyngeal nerves and the cranial cervical ganglion form a pharyngeal plexus which innervates the caudal pharyngeal muscles and the cranial esophagus. The **cranial laryngeal nerve** leaves the vagus nerve at the distal ganglion and passes ventrally to the larynx, where it supplies the cricothyroid muscle and the laryngeal mucosa. The origin of the recurrent laryngeal nerve was seen in the thoracic inlet. It can now be found terminating as the **caudal laryngeal nerve** which enters the larynx under the caudal edge of the cricopharyngeus muscle. It innervates all the muscles of the larynx except the cricothyroideus.

Locate the sympathetic trunk where it is joined with the vagus. Follow it cranially to a level ventral and medial to the tympanic bulla, where the two trunks separate. Remove the origin of the digastricus to observe these nerves. The sympathetic trunk is ventral to the vagus. The

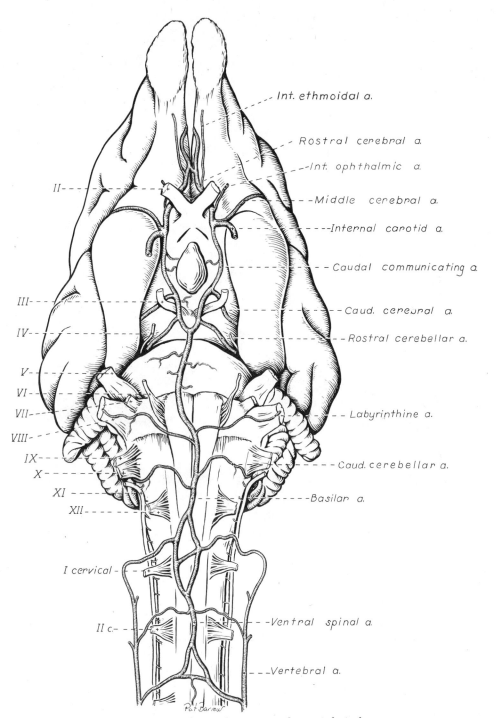

Figure 189a. *Cranial nerves and arterial circle.*

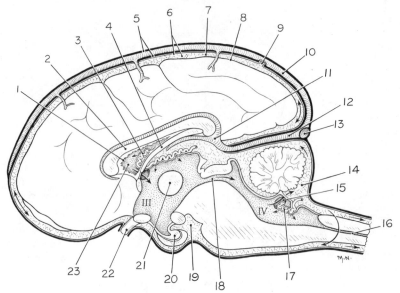

Figure 189b. *Meninges and ventricles.*

1. *Cut edge of septum pellucidum*
2. *Corpus callosum*
3. *Choroid plexus*
4. *Fornix of hippocampus*
5. *Dura*
6. *Arachnoid membrane and trabeculae*
7. *Subarachnoid space*
8. *Pia*
9. *Arachnoid villus*
10. *Dorsal sagittal sinus*
11. *Great cerebral vein*
12. *Straight sinus*

13. *Transverse sinus*
14. *Cerebellomedullary cistern*
15. *Lateral aperture of fourth ventricle*
16. *Central canal*
17. *Choroid plexus*
18. *Mesencephalic aqueduct*
19. *Intercrural cistern*
20. *Hypophysis*
21. *Interthalamic adhesion*
22. *Optic nerve*
23. *Lateral ventricle*

distal ganglion of the vagus is located dorsal to this separation. Trace the sympathetic trunk cranial to the separation and note the **cranial cervical ganglion.** This is the most cranial group of cell bodies of sympathetic postganglionic axons. On the lateral side observe the internal carotid artery coursing dorsocranially over the lateral surface of the cranial cervical ganglion. Notice the dense plexus which these nerves form in the immediate vicinity of the ganglion.

11. The **accessory** or **eleventh cranial nerve** passes through the jugular foramen and the occipitotympanic fissure along with the ninth and tenth cranial nerves. It passes caudally and innervates the sterno-mastoideus and cleidomastoideus muscles. Its caudodorsal course to supply the cleidocervicalis, the omotransversarius and the trapezius has been dissected.

12. The **hypoglossal** or **twelfth cranial nerve** passes through the hypoglossal canal. It passes ventrorostrally lateral to the vagosympathetic trunk and the carotid arteries and deep to the mandible. Deep to the mylohyoideus it innervates the extrinsic and intrinsic muscles of the tongue.

THE NERVOUS SYSTEM

The nervous system may be divided into the **central nervous system,** consisting of the brain and spinal cord, and the **peripheral nervous system,** composed of cranial, spinal and peripheral nerves.

CEREBRAL MENINGES

The brain and spinal cord are covered by three membranes of connective tissue, the **meninges.** The **dura mater,** or pachymeninx, is the thickest of these and the most external. Throughout most of the vertebral canal the dura is separated from the periosteum of the bony canal by the loose connective tissue of the epidural space.

As the spinal cord approaches the brainstem the dura adheres to the periosteum in the first two cervical vertebrae and to the atlanto-occipital membranes. Inside the cranial cavity the dura and periosteum are fused. Free the hemisectioned brain from the skull of the sagittally split head starting at the dorsal margin. Remove the dura which adheres to the inside of the frontal, parietal and temporal bones. On one-half of the head a fold of dura will be found extending ventrally from the midline in the longitudinal cerebral fissure between the two cerebral hemispheres. This is the **falx cerebri,** which must be removed to allow reflection of the enclosed cerebral hemisphere.

The **pia mater** and **arachnoid** (the **leptomeninges**) are the other two connective tissue coverings of the central nervous system. The pia mater adheres to the external surface of the nervous tissue. The arachnoid lies between the pia and the dura and sends delicate trabeculae to the pia which closely invest the blood vessels that course on the surface of the pia. The space between the pia and the arachnoid is the subarachnoid space, which is filled with cerebrospinal fluid.

Subarachnoid cisterns occur in areas where the arachnoid and pia are separated. The largest cistern is the **cerebellomedullary cistern** located in the angle between the cerebellum and medulla. Cerebrospinal fluid may be obtained from this cistern.

ARTERIES

Examine the arterial supply to the brain on the latex-injected specimen. The arteries to the cerebrum and cerebellum are branches from the vessels on the ventral surface of the brain. The **basilar artery** is formed by the terminal branches of the **vertebral arteries** which enter the floor of the vertebral canal through the lateral vertebral foramina of the atlas. It is continuous caudally with the **ventral spinal artery** of the spinal cord. The basilar artery courses along the midline of the ventral surface of the medulla and pons and then divides into two branches that form the caudal portion of the arterial circle of the brain. (See Fig. 189.)

The **internal carotid arteries** are the other main source of blood to the arterial circle of the brain. Entering the middle fossa of the cranial cavity from the foramen lacerum, each divides into a middle cerebral, rostral cerebral and caudal communicating artery. The small **caudal communicating arteries** course caudally and join the terminal branches of the basilar artery. Rostrally the two rostral cerebral arteries anastomose, completing the arterial circle on the ventral surface of the brain.

The **cerebral arterial circle** surrounds the pituitary gland, which receives small branches from the circle as well as directly from the internal carotid artery. Put the two halves of the latex-injected brain together to observe this arterial circle. Using the hemisectioned latex-injected brain identify and trace the following vessels.

The **rostral cerebral artery** is a terminal branch of the internal carotid artery at the rostral aspect of the circle. It courses dorsally lateral to the optic chiasm and continues between the two frontal lobes in the longitudinal fissure. It supplies the medial surface of the rostral half of the cerebral hemisphere.

The **middle cerebral artery** arises from the arterial circle at the level of the rostral aspect of the pituitary gland. It courses laterally rostral to the piriform lobe on the ventral surface of the olfactory peduncle. It continues dorsolaterally over the cerebral hemispheres, where it branches to supply the lateral surface.

The **caudal cerebral artery** arises from the caudal communicating artery at the level of the caudal aspect of the pituitary gland, rostral to the oculomotor nerve. The artery courses caudodorsally, following the optic tract over the lateral aspect of the thalamus to the longitudinal fissure. It passes rostrally on the corpus callosum to supply the medial surface of the caudal portion of the cerebral hemisphere. It also supplies the diencephalon and the rostral mesencephalon.

The **rostral cerebellar artery** leaves the caudal third of the arterial circle and courses dorsocaudally along the pons and the middle cerebellar peduncle to the lateral cerebellar hemisphere. It supplies the caudal midbrain and the rostral half of the cerebellum.

The **labyrinthine artery** leaves the basilar artery at the level of the trapezoid body and courses laterally with the facial and vestibulocochlear nerves through the internal acoustic meatus to supply the inner ear.

The **caudal cerebellar artery** is a branch of the basilar artery near the middle of the medulla. It courses dorsally to supply the caudal portion of the cerebellum.

VEINS

The **venous sinuses** of the cranial dura mater are venous passageways located within the dura or within bony canals in the skull. These sinuses receive the veins draining the brain and the bones of the skull. They convey venous blood to the paired maxillary, internal jugular and vertebral veins and to the vertebral venous plexuses. The following venous sinuses should be located.

The **dorsal sagittal sinus** is located in the attached edge of the falx cerebri, which is a fold of dura extending ventrally into the longitudinal fissure between the two cerebral hemispheres. Caudally this sinus enters the foramen for the dorsal sagittal sinus in the occipital bone where the tentorium attaches. There it joins the right and left transverse sinuses.

Each **transverse sinus** runs laterally through the transverse canal and sulcus. At the distal end of the sulcus at the dorsal border of the petrosal bone the sinus divides into a temporal and a sigmoid sinus. The **temporal sinus** courses caudolateral to the petrosal bone to the retroarticular foramen, where it emerges and joins the maxillary vein.

The **sigmoid sinus** forms an S-shaped curve as it courses over the dorsomedial side of the petrous temporal bone. It passes through the jugular foramen and emerges from the occipitotympanic fissure. Within the fissure the ventral petrosal sinus enters rostrally and the vertebral and internal jugular veins leave caudally. The vertebral vein descends the neck through the transverse foramina of the cervical vertebrae. The internal jugular vein was seen previously in the carotid sheath. A branch of the sigmoid sinus is continuous caudally through the condyloid canal with the vertebral venous plexuses in the vertebral canal.

The paired **cavernous sinus** lies on the floor of the middle cranial fossa from the orbital fissures to the petro-occipital canals. Emissary veins connect each cavernous sinus with the orbital plexus of veins rostrally and with the maxillary vein laterally. These sinuses are continued caudally by the ventral petrosal sinus, which lies in the petro-occipital canal. Two or three intercavernous sinuses connect the left and right cavernous sinuses rostral and caudal to the pituitary gland.

The **vertebral venous plexuses** are paired vessels which lie on the floor of the vertebral canal in the epidural connective tissue. They extend from the venous sinuses of the cranium throughout the vertebral canal and will be seen later when the spinal cord is removed from the vertebral canal.

BRAIN

The brain is composed of the embryologically segmented brainstem and two suprasegmental portions, the cerebrum (telencephalon) and the cerebellum (dorsal metencephalon). The **brainstem** includes the myelen-cephalon (medulla), the ventral metencephalon (pons), the mesencephalon (midbrain) and the diencephalon (interbrain—epithalamus, thalamus and hypothalamus).

Dissect and identify the following structures on the intact brain that has been provided.

CEREBRUM—SURFACE STRUCTURES

The cerebrum is divided into two cerebral hemispheres by the **longi-tudinal fissure.** Each cerebral hemisphere has outward folds (convolu-tions) called **gyri** and inward folds called **sulci.** Identify the following gyri and sulci (Figs. 190, 191): the rostral and caudal parts of the lateral rhinal sulci; the pseudosylvian fissure; the rostral and caudal sylvian gyri; the cruciate sulcus; the postcruciate and precruciate gyri; the coro-nal sulcus; the marginal gyrus and sulcus; the ectomarginal gyrus.

Each cerebral hemisphere may be divided into lobes which are named by the bone of the skull which overlies the area. The **frontal lobe** is that portion of each cerebral hemisphere rostral to the cruciate sulcus. The precruciate gyrus is part of this lobe and functions as part of the motor cortex. The **parietal lobe** is caudal to the cruciate sulcus and dorsal to the sylvian gyri, and extends caudally approximately to the caudal one-third of the cerebral hemisphere. The postcruciate and rostral ectomarginal gyri are found in this lobe and function as part of the motor and somes-thetic sensory cerebral cortex. The **occipital lobe** includes the caudal one-third of the cerebral hemisphere. Caudal portions of this lobe on both medial and lateral sides function as the visual cortex. The **temporal lobe** is composed of the gyri and sulci on the ventrolateral aspect of the

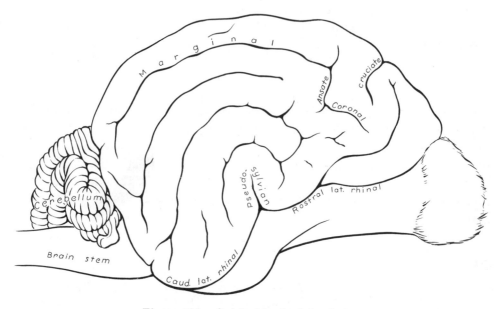

Figure 190.   Sulci of brain, lateral view.

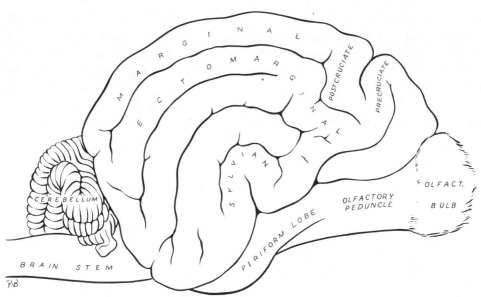

Figure 191. *Gyri of brain, lateral view.*

cerebral hemisphere. Parts of the sylvian gyri are located here and function as the auditory cortex.

The **rhinal sulcus** separates the phylogenetically new cerebrum, neopallium, above from the older olfactory cerebrum, paleopallium, below. Portions of the paleopallium that are visible are the **olfactory bulb,** which rests on the cribriform plate, and the **olfactory peduncle,** which joins the bulb to the cerebral hemisphere. The **olfactory peduncle** courses caudally with the **olfactory tract** on its ventral surface. Caudally the olfactory tract divides into **lateral** and **medial olfactory striae.** Observe the lateral olfactory stria passing caudally to the **piriform lobe,** which forms a ventral bulge just lateral to the pituitary gland and medial to the temporal lobe of the neopallium.

Each gyrus contains gray matter superficially and white matter in its center. The gray matter or cerebral cortex is composed of six layers of neuronal cell bodies. The white matter or medullary core contains the processes of neurons coursing to and from the overlying cortex.

CEREBELLUM

The **cerebellum** is derived from the dorsal portion of the metencephalon and lies caudal to the cerebrum and dorsal to the fourth ventricle. The **transverse cerebral fissure** separates it from the cerebrum. It is connected to the brainstem by three pairs of cerebellar peduncles and by portions of the roof of the fourth ventricle.

The **choroid plexus** is a compact mass of pia and blood vessels which protrudes into the lumen of the fourth ventricle. It is visible caudolateral to the cerebellum.

Identify the **transverse fibers of the pons** on the ventral surface of the brainstem. Follow these fibers laterally as they course dorsocaudally into the cerebellum on each side as the **middle cerebellar peduncle.** At the point where they merge into the cerebellum cut this peduncle with a scalpel. Continue the cut rostral and caudal to the middle cerebellar peduncle and detach the cerebellum from the pons on that side. This will sever the **rostral cerebellar peduncle,** which attaches rostrally, and the **caudal cerebellar peduncle,** which attaches caudally. Cut these peduncles on the opposite side and remove the cerebellum. The rostral cerebellar peduncle contains mainly efferent neurons to the cerebellum. Afferent to the cerebellum neurons pass primarily through the middle and caudal cerebellar peduncles.

The cerebellum is composed of lateral **cerebellar hemispheres** and a middle portion, the **vermis.** The gyri of the cerebellum are known as **folia.** These are grouped into cerebellar lobules that have specific names. The vermis, subdivided into a number of lobules, comprises the entire middle portion of the cerebellum which is directly above the fourth ventricle. Some of its lobules are found on the ventral surface of the cerebellum facing the fourth ventricle. Each lateral hemisphere projects over the cerebellar peduncles and the adjacent brainstem.

Make a median incision through the vermis, hemisectioning the cerebellum. Examine the cut surface. Note the pattern of white matter as it branches and arborizes from the medulla into the folia. Observe the medullary rays of white matter and the cerebellar cortex.

BRAINSTEM — SURFACE STRUCTURES

Diencephalon. Examine the ventral surface of the brainstem and locate the following structures.

The optic or second cranial nerves form the **optic chiasm** of the diencephalon rostral to the pituitary gland. The **optic tracts** course dorsally from the chiasm and will later be seen to pass over the lateral surface of the diencephalon to enter the thalamus.

Caudal to the optic chiasm on the midline is the **hypophysis,** attached by the **infundibulum** to the **tuber cinereum** of the hypothalamus. If the gland is missing the lumen of the infundibulum will be evident. This communicates with the overlying third ventricle of the diencephalon.

The **mamillary bodies** of the hypothalamus bulge ventrally caudal to the tuber cinereum. They demarcate the most caudal extent of the hypothalamus on the ventral surface of the brainstem.

Mesencephalon. Between the mamillary bodies of the hypothalamus and the transverse fibers of the pons is the ventral surface of the **mesencephalon** (midbrain). Descending tracts that connect portions of the cerebral cortex with lower brainstem centers and the spinal cord course on the ventral surface of the midbrain. These are grouped together on each side as the **crus cerebri.** The oculomotor or third cranial nerve leaves the midbrain medial to the crus.

The mesencephalic structures dorsal to the mesencephalic aqueduct

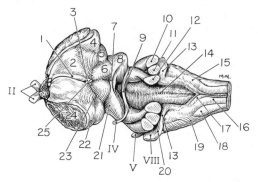

Figure 192. *Dorsolateral view of brainstem.*

1. Stria habenularis thalami
2. Thalamus
3. Habenular commissure
4. Lateral geniculate nucleus
5. Medial geniculate nucleus
6. Rostral colliculus
7. Commissure of caudal colliculus
8. Caudal colliculus
9. Crossing of trochlear nerve fibers in rostral medullary velum
10. Middle cerebellar peduncle
11. Caudal cerebellar peduncle
12. Rostral cerebellar peduncle
13. Dorsal cochlear nucleus
14. Median sulcus in fourth ventricle
15. Lateral cuneate nucleus
16. Fasciculus cuneatus
17. Fasciculus gracilis
18. Spinal tract of trigeminal nerve
19. Superficial arcuate fibers
20. Left ventral cochlear nucleus
21. Brachium of caudal colliculus
22. Optic tract
23. Brachium of rostral colliculus
24. Cut surface between cerebrum and brainstem
25. Pineal body
II. Optic nerves
IV. Trochlear nerve
V. Trigeminal nerve
VIII. Vestibulocochlear nerve

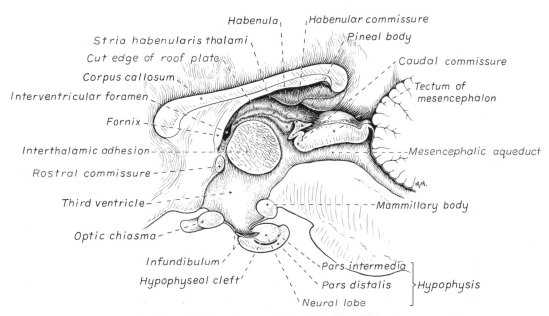

Figure 193. *Diencephalon, median section.*

comprise the **tectum** of the midbrain. Four dorsal bulges, the **corpora quadrigemina,** are evident. The rostral pair are the **rostral colliculi** that function with the visual system. The caudal pair are the **caudal colliculi** that function with the auditory system.

The trochlear or fourth cranial nerve courses laterally out of the roof of the fourth ventricle adjacent to the caudal colliculus. It continues rostroventrally on the lateral surface of the midbrain.

The **lateral lemniscus** is on the lateral side of the midbrain coursing rostrodorsally to the caudal colliculus. The **brachium of the caudal colliculus** runs rostroventrally from the caudal colliculus to the medial geniculate body of the thalamus. On the dorsal surface the **commissure of the caudal colliculi** can be seen crossing between these two structures. The rostral colliculus is connected to the lateral geniculate body of the thalamus.

Ventral Metencephalon.   The metencephalic portion of the rhombencephalon includes a segment of the brainstem, the **pons,** and a dorsal development, the **cerebellum.** The ventral surface of the pons includes the transverse fibers of the pons rostrally and extends through the **trapezoid body** caudally.

A portion of the descending fibers of the crus cerebri of the midbrain are continued through the pons dorsal to its transverse fibers and appear on the ventral surface of the pons coursing caudally over the surface of the trapezoid body. These are the **pyramidal tracts,** which are separated from each other by the **ventral median fissure.** They continue caudally on the medulla.

The abducens or sixth cranial nerves leave the pons through the trapezoid body on the lateral border of each pyramidal tract.

On the lateral side of the pons, cranial nerves 5, 7 and 8 are located. Cranial nerve 5, the trigeminal nerve, runs rostroventrally from the groove between the middle cerebellar peduncle rostrally and the trapezoid body caudally.

Cranial nerve 7, the facial nerve, leaves the lateral surface of the pons through the trapezoid body caudal to the trigeminal nerve and rostroventral to the eighth cranial nerve.

Cranial nerve 8, the vestibulocochlear nerve, is on the lateral side of the pons at the most lateral extent of the trapezoid body. Part of this nerve contributes fibers to the trapezoid body; part enters the pons directly; and part courses over the dorsal surface of the pons and caudal cerebellar peduncle just caudal to where the latter passes dorsally into the cerebellum. The **cochlear nuclei** are located in this nerve as it courses over the lateral and dorsal surfaces of the pons.

Examine the dorsal surface of the pons (Fig. 192). On either side of the fourth ventricle are the cut ends of the three cerebellar peduncles. The rostral cerebellar peduncle is medial and courses rostrally into the mesencephalon; the middle cerebellar peduncle is lateral and enters from the lateral side of the pons; and the caudal cerebellar peduncle is in the middle, entering from the myelencephalon.

The **rostral medullary velum** forms the roof of the fourth ventricle between the caudal colliculi of the mesencephalon rostrally and the ventral surface of the cerebellum caudally. The crossing fibers of the trochlear nerves course through this velum. The velum in the preserved specimen lies on the floor of the fourth ventricle and covers the opening of the mesencephalic aqueduct. Insert a probe under the caudal cut edge and raise the velum to demonstrate its attachments.

The groove in the center of the floor of the fourth ventricle is the **median sulcus.** On the lateral wall the longitudinal groove is the sulcus limitans. Just lateral to this latter groove at the level of the entering eighth nerve fibers there is a slight dorsal bulge of the pons. This demarcates the location of the **vestibular nuclei.**

Myelencephalon (Medulla). On the ventral surface of the myelencephalon, the **pyramidal tracts** and **ventral median fissure** can be followed caudally until the fissure is obliterated over a short distance by the crossing of the fibers in the pyramidal tracts. This is the **decussation of the pyramids** and is located at the level of the emerging fibers of the twelfth cranial nerve, the hypoglossal nerve. The hypoglossal nerve emerges as a number of fine rootlets from the ventrolateral side of the myelencephalon caudal to the trapezoid body. Observe that these fibers are in the same sagittal plane as the third and sixth cranial nerves rostrally and the ventral rootlets of the spinal cord caudally. The junction of the myelencephalon and the spinal cord is arbitrarily placed between the hypoglossal fibers and the ventral rootlets of the first cervical spinal nerve.

Dorsolateral to the emergence of the hypoglossal fibers and the fibers of the ventral root of the first cervical spinal nerve, a nerve is located running lengthwise along the lateral surface of the spinal cord. This is the eleventh cranial nerve, the accessory nerve. Its spinal rootlets emerge from the lateral surface of the spinal cord as far caudally as the seventh cervical segment. They emerge between the level of the dorsal and ventral rootlets of the cervical spinal nerves and course cranially through the vertebral canal and foramen magnum. The accessory leaves the cranial cavity through the jugular foramen with the ninth and tenth cranial nerves.

Cranial nerves 9 and 10, the glossopharyngeal and vagus nerves, leave the lateral side of the rostral end of the myelencephalon caudal to the eighth cranial nerve and rostral to the accessory nerve. These rootlets are small and are rarely preserved on the brain when it is removed.

Examine the dorsal surface of the myelencephalon caudal to the fourth ventricle. The median groove is the **dorsal median sulcus.** The narrow longitudinal bulge flanking this sulcus is the **fasciculus gracilis.** This longitudinal tract ascends the entire length of the spinal cord in this position. Here at the caudal end of the myelencephalon it ends in the **nucleus gracilis.** This nucleus is located where the fasciculus ends, at the caudal end of the fourth ventricle.

The groove lateral to the fasciculus gracilis is the **dorsal intermediate sulcus.** The longitudinal bulge lateral to this is the **fasciculus cuneatus.** This tract also ascends the dorsal aspect of the spinal cord, starting in the midthoracic region. The fasciculus cuneatus diverges laterally at the caudal end of the fourth ventricle and ends in a slight bulge. This bulge represents the **lateral cuneate nucleus** and is known as the **cuneate tubercle.** Rostrally the lateral cuneate nucleus is continuous with the caudal cerebellar peduncle.

The groove on the caudodorsal surface of the myelencephalon lateral to the fasciculus cuneatus is the **dorsolateral sulcus.** The longitudinal bulge lateral to this is the **spinal tract of the trigeminal nerve.** This tract emerges on the lateral surface of the myelencephalon caudal to a band of obliquely-ascending fibers, the superficial arcuate fibers. It courses caudally to the level of the first cervical segment of the spinal cord. The dorsal rootlets of the spinal cord enter through the dorsolateral sulcus along the spinal cord.

The roof of the fourth ventricle caudal to the cerebellum is the **caudal medullary velum.** It is a thin layer composed of ependyma lining the ventricle and a supporting layer of vascularized pia. It attaches to the cerebellum rostrally and to the caudal cerebellar peduncle and the fasciculus gracilis laterally and caudally. Its attachment caudally at the apex is known as the **obex.** At this level the fourth ventricle is continuous with the central canal of the spinal cord.

At the level of the eighth cranial nerve there is an opening in the caudal medullary velum known as the **lateral aperture of the fourth ventricle.** The cerebrospinal fluid produced in the ventricular system communicates with the subarachnoid space of the meninges via this aperture. It then courses through the subarachnoid space over the entire surface of the brain and spinal cord and is absorbed into the venous system. Most of this absorption occurs where the arachnoid is in close apposition to the cerebral venous sinuses. Cerebrospinal fluid is also absorbed from the subarachnoid space, where the spinal nerves leave the vertebral canal through the intervertebral foramina, and along the olfactory and optic nerves.

The **choroid plexus of the fourth ventricle** bulges into the lumen of the ventricle and extends through the lateral aperture, where it was seen caudal to the cerebellum before the latter was removed. The choroid plexus consists of blood vessels covered by a layer of pia and ependyma. Cerebrospinal fluid is produced by the choroid plexuses of the fourth, third and lateral ventricles.

TELENCEPHALON (CEREBRUM)

Separate the two cerebral hemispheres at the longitudinal fissure. Expose the band of fibers that course transversely from one hemisphere to the other in the depth of the fissure. This structure is the **corpus callosum.** It is divided into a rostral genu, a middle body and a caudal splenium.

Separate the telencephalon from the diencephalon on the left side by the following procedure. Completely divide the corpus callosum longitudinally along the midline in the depth of the longitudinal fissure. Continue the cut rostrally through the genu and ventrally through the rostral

Plate I. Telencephalon

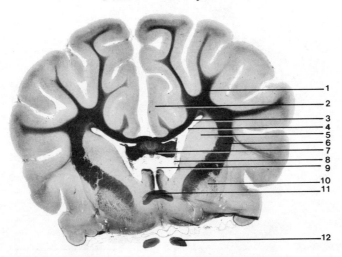

1. *Corona radiata*
2. *Cingulate gyrus*
3. *Corpus callosum*
4. *Lateral ventricle*
5. *Caudate nucleus*
6. *Internal capsule*

7. *Column of fornix*
8. *Interventricular foramen*
9. *Third ventricle*
10. *Lentiform nucleus*
11. *Rostral commissure*
12. *Optic nerve*

Plate II. Diencephalon

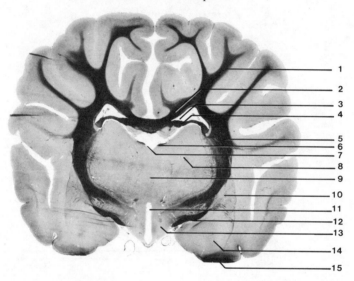

1. Corona radiata
2. Corpus callosum
3. Lateral ventricle
4. Fornix
5. Internal capsule
6. Stria habenularis
7. Third ventricle
8. Thalamus
9. Interthalamic adhesion
10. Lentiform nucleus
11. Third ventricle
12. Optic tract
13. Hypothalamus
14. Amygdala
15. Pyriform lobe

Plate III. *Mesencephalon*

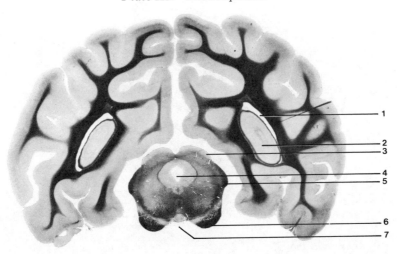

1. *Lateral ventricle*
2. *Hippocampus*
3. *Rostral colliculus*
4. *Mesencephalic aqueduct*
5. *Brachium of caudal colliculus*
6. *Crus cerebri*
7. *Intercrural fossa*

Plate IV. Cerebellum and Myelencephalon

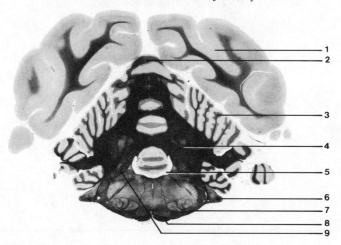

1. Occipital lobe
2. Folium of cerebellar vermis
3. Folium of lateral cerebellar hemisphere
4. Medulla of cerebellum
5. Cerebellar peduncle
6. Fourth ventricle
7. Vestibulocochlear nerve and cochlear nucleus
8. Trapezoid body
9. Pyramid

commissure just dorsal to the optic chiasm. On the ventral surface follow the optic tract in a dorsocaudal direction and cut the fibers rostral to this tract which attach the cerebral hemisphere to the brainstem. Continue this separation over the dorsal aspect of the diencephalon. Cut any remaining attachments and remove the hemisphere from the diencephalon.

See Figure 194 to determine the structures that will remain after the left cerebral hemisphere is removed. The mass of fibers that have been cut rostrodorsolateral to the optic tract is the **internal capsule.** This capsule contains projection fibers that course between the telencephalon and the diencephalon and descend from the telencephalon to the brainstem and the spinal cord. The fibers that were previously seen in the

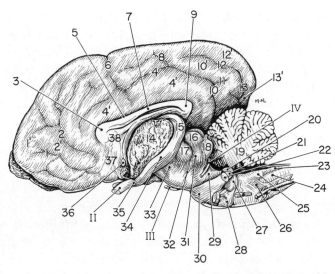

Figure 194. *Medial surface of right cerebrum and lateral surface of brainstem.*

1 Ectogenual sulcus	*18 Caudal colliculus*
1' Proreal gyrus	*19 Arbor vitae cerebelli*
2 Genual sulcus	*20 Rostral cerebellar peduncle*
2' Genual gyrus	*21 Caudal cerebellar peduncle*
3 Genu of corpus callosum	*22 Middle cerebellar peduncle*
4 Splenial sulcus	*23 Fasciculus cuneatus*
4' Cingulate gyrus	*24 Spinal tract of trigeminal nerve*
5 Callosal sulcus of corpus callosum	*25 Lateral cuneate nucleus*
6 Cruciate sulcus	*26 Superficial arcuate fibers*
7 Body of corpus callosum	*27 Cochlear nuclei*
8 Lesser cruciate sulcus	*28 Trapezoid body*
9 Splenium of corpus callosum	*29 Lateral lemniscus*
10 Splenial sulcus	*30 Transverse fibers of pons*
10' Splenial gyrus	*31 Brachium of caudal colliculus*
11 Caudal horizontal ramus of calcarine sulcus	*32 Transverse crural tract*
12 Suprasplenial sulcus	*33 Crus cerebri*
12' Occipital gyrus	*34 Left optic tract*
13 Postsplenial sulcus	*35 Optic chiasm*
13' Postsplenial gyrus	*36 Rostral commissure*
14 Cut surface between cerebrum and brainstem	*37 Paraterminal gyrus*
15 Lateral geniculate nucleus	*38 Septum pellucidum*
16 Rostral colliculus	*II Optic nerve*
17 Medial geniculate nucleus	*III Oculomotor nerve*
	IV Trochlear nerve

mesencephalon as the crus cerebri descend through the internal capsule.

Notice how the optic tract curves around the caudal edge of the internal capsule and ascends dorsocaudally to enter the thalamus of the diencephalon.

Examine the isolated left cerebral hemisphere. On its medial surface note the thin sheet of tissue ventral to the corpus callosum. This is the **septum pellucidum,** which is more developed rostrally where it extends from the genu to the rostral commissure. A thickening in the septum dorsal and rostral to the rostral commissure represents the **septal nuclei.** Caudal to the rostral commissure the septum attaches to a column of fibers that course rostrally and then descend in a rostroventral curve behind the rostral commissure. These fibers are part of the **fornix.** They connect the hippocampus with the diencephalon and rostral cerebrum. Dissect away the septum pellucidum, exposing the fornix. The fornix begins caudally by the accumulation of fibers on the lateral side of the hippocampus. These form the **crus** of the fornix. The crura join rostral to the hippocampus, dorsal to the thalamus, to form the **body** of the fornix.

The body of the fornix courses rostrally and then descends rostroventrally as the **column** of the fornix. At the rostral commissure some fibers course dorsal and rostral to the rostral commissure, but most descend caudal to the commissure and continue caudoventrally lateral to the third ventricle to reach the mamillary body of the hypothalamus. This descending column may be more evident on the intact right cerebral hemisphere.

The curved cavity exposed lateral to the septum pellucidum and ventral to the corpus callosum is the **lateral ventricle.** It communicates with the third ventricle of the diencephalon by the **interventricular foramen,** which is located caudal to the column of the fornix at the level of the rostral commissure.

Locate the caudal part of the lateral rhinal sulcus in the left temporal lobe. Cut through this sulcus into the lateral ventricle. The smooth, curved bulge exposed in the wall of the ventricle here is the caudal surface of the hippocampus as it ascends in a dorsorostral curve. To remove the hippocampus intact from the ventricle cut the column of the fornix dorsal to the rostral commissure. Grasp the fornix with the forceps and, with the blunt end of the scalpel, gently roll the hippocampus out of the lateral ventricle. Its attachment ventrally in the temporal horn may be cut to completely free the hippocampus.

The **hippocampus** phylogenetically belongs to the archipallium and is an internal gyrus of the telencephalon which has been rolled into the lateral ventricle by the lateral expansion of the neopallium. Notice that the hippocampus begins ventrally in the temporal lobe and curves, first caudodorsally and then rostrodorsally, over the diencephalon to reach its caudodorsal aspect. At that point the hippocampus ends and its fibers continue rostrally as the body and the column of the fornix. Notice the crus of the fornix on its lateral surface. Place the previously removed hippocampus over the exposed left diencephalon to see its normal relationship with that structure.

Attached to the lateral free edge of the crus of the fornix is a network of blood vessels covered by meninges. This is the **choroid plexus of the lateral ventricle.** Its anatomical structure is the same as the plexus of the fourth ventricle. The surface of the choroid plexus that faces the lumen of the ventricle is the ependymal layer derived from the embryonic neural tube. This layer of ependyma is attached on one side to the free edge of the fornix. In order to complete the wall of the lumen of the lateral ventricle it must be attached on its other side. This attachment is in the groove between the thalamus and the caudate nucleus. This layer of ependyma from the groove to the fornix forms part of the medial wall of the lateral ventricle. Branches of the middle and caudal cerebral arteries covered by pia push this layer of ependyma into the lumen of the lateral ventricle. The result is the formation of the choroid plexus.

Examine the **rostral commissure.** On each side, this commissure connects rostrally to the olfactory peduncles and caudally to the piriform lobes.

Examine the floor of the lateral ventricle of the removed left hemisphere. The bulge which enlarges rostrally is the **caudate nucleus.** This is one of the subcortical nuclei of the telencephalon, a part of the corpus striatum. Its rostral extremity is the **head.** Caudal to this the **body** rapidly narrows into a small **tail,** which courses over the internal capsule fibers in the floor of the ventricle and continues, first caudoventrally, then rostroventrally to end in the piriform lobe, where the hippocampus attaches. With the blunt end of a scalpel free the caudate nucleus from the medial side of the internal capsule.

Dorsolateral to the caudate nucleus the internal capsule forms the lateral wall of the lateral ventricle. At the dorsolateral angle of the lateral ventricle the fibers of the internal capsule meet those of the corpus callosum. This interdigitation of fibers is known as the **corona radiata,** which radiates in all directions to reach the gray matter of the cerebral cortex.

Remove the pia and arachnoid from the surface of the right cerebral hemisphere. Expose the corona radiata by removing the gray matter with the scalpel handle. (Following fixation the brain specimen was frozen and then thawed. This procedure dissociates cell bodies from fiber tracts and makes it possible to remove the gray matter without damaging the corona radiata.)

Begin this removal of gray matter on the medial side of the hemisphere. The cingulate gyrus is located dorsal to the corpus callosum. Remove the gray matter of the cingulate gyrus to expose its fibers, which form the **cingulum.** Many of the fibers in the cingulum are long association fibers that course longitudinally from one end of the hemisphere to the other. Demonstrate this by freeing some fibers rostral to the genu of the corpus callosum and by stripping them caudally. Remove the cingulum and demonstrate the transverse course of the fibers of the corpus callosum by stripping these from their cut edge towards the hemisphere.

Remove the gray matter from the gyri on the lateral surface of the rostral half of the hemisphere. Examine the white matter of the gyri.

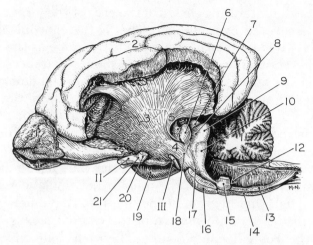

Figure 195. *Lateral view of brain showing projection pathways.*

1. *Olfactory bulbs*
2. *Left cerebral hemisphere*
3. *Internal capsule (lateral view)*
4. *Crus cerebri*
5. *Acoustic radiation*
6. *Medial geniculate nucleus*
7. *Rostral colliculus*
8. *Brachium of caudal colliculus*
9. *Caudal colliculus*
10. *Lateral lemniscus*
11. *Cerebellum*
12. *Location of dorsal nucleus of trapezoid body*
13. *Location of olivary nucleus*
14. *Pyramid*
15. *Trapezoid body*
16. *Transverse fibers of pons*
17. *Pyramidal and corticopontine tracts (longitudinal fibers of pons)*
18. *Transverse crural tract*
19. *Piriform lobe*
20. *Optic tract (cut to show internal capsule)*
21. *Optic chiasm*
II. *Optic nerve*
III. *Oculomotor nerve*

Short association fibers, the arcuate fibers, course between adjacent gyri.

The internal capsule is situated lateral to the caudate nucleus. Expose the lateral surface of the internal capsule (Fig. 195).

When you have completed the dissection of the right cerebral hemisphere, all that remains is the corpus callosum, the internal capsule and part of the corona radiata.

DIENCEPHALON

Remove the dorsal portion of the right cerebral hemisphere including the corpus callosum. This will expose the lateral ventricle, with the caudate nucleus on the floor rostrally and the hippocampus on the floor caudally. The internal capsule forms the lateral wall.

Examine the diencephalon. The internal capsule bounds it laterally. The thalamus and epithalamus can be seen on the dorsal aspect of the diencephalon (Figs. 192, 193).

Three structures comprise the epithalamus. The **stria habenularis** lies on either side of the midline, coursing from the rostroventral aspect of the hypothalamus dorsally and caudally over the thalamus to the dorsocaudal aspect of the diencephalon. Here the stria enters the **habenular nucleus.** Caudal to the habenular nucleus is the small, unpaired **pineal body.** This caudal projection from the diencephalon is small in the dog but very prominent in larger domestic animals.

A space can usually be found between the stria habenularis of each side. This is part of the **third ventricle.** It is covered by a thin remnant of the roof plate of the neural tube, a layer of ependyma which extends from one stria habenularis to the other. Branches of the caudal cerebral artery course over the diencephalon and form the **choroid plexus of the third ventricle.**

The thalamus lies between the stria habenularis medially and the internal capsule laterally. An eminence on the caudodorsal surface of the thalamus is the **lateral geniculate body,** which receives fibers of the optic tract and functions in the visual system. The lateral geniculate body is connected with the rostral colliculus. Caudoventral to the lateral geniculate body is the **medial geniculate body** of the thalamus. This nucleus functions in the auditory system and is connected to the caudal colliculus by the brachium of the caudal colliculus.

In the third ventricle between the stria habenularis of each side observe the **interthalamic adhesion** between the right and left thalami.

Make a median section of the brainstem from the optic chiasm through the medulla. Observe the interthalamic adhesion and note the smooth surface of the third ventricle surrounding it. The ventricle is bounded dorsally by the roof plate between each stria habenularis. It connects rostrally with each lateral ventricle through the interventricular foramen, which is caudal to the column of the fornix and dorsal to the rostral commissure. Caudally it is continuous with the mesencephalic aqueduct. Rostroventral to the interventricular foramen the ventricle is bounded by the lamina terminalis. The **lamina terminalis** demarcates the most rostral extent of the embryonic neural tube. (From this point each cerebral hemisphere developed laterally.) The third ventricle extends into the infundibulum of the pituitary gland as the **infundibular recess.**

The hypothalamus forms the ventral portion of the diencephalon and the lateral wall of the third ventricle on each side. On its ventral surface are the optic chiasm, the tuber cinereum and the pituitary. Caudal to the tuber cinereum are the mamillary bodies of the hypothalamus.

SPINAL CORD

Before exposing the spinal cord study the ligaments associated with the vertebral arches and their spines. Remove the epaxial muscles from the spines of the vertebrae. The **nuchal ligament** is a bundle of longitudinal, yellow, elastic fibers that attaches cranially to the spine of the axis. It extends caudally through the cervical epaxial muscles to the spine of the first thoracic vertebra.

The **supraspinous ligament** extends from the spinous process of the

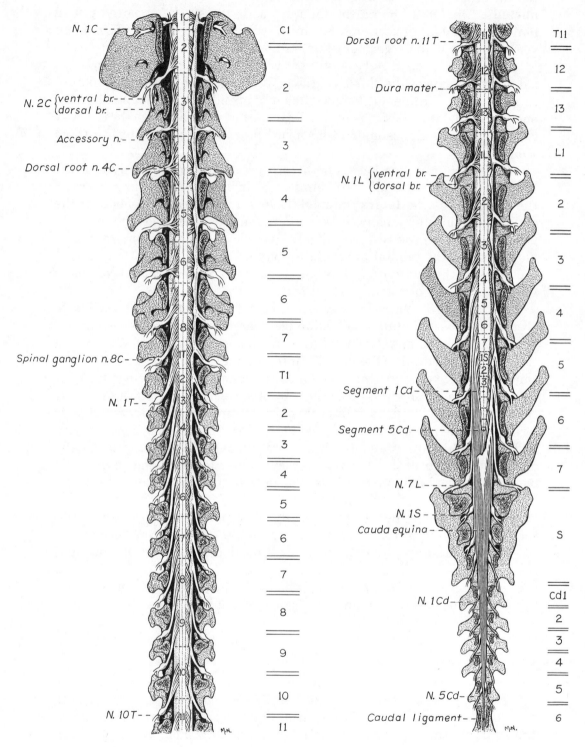

Figure 196. Dorsal roots of spinal nerves and spinal cord segments. (Figures on the right represent levels of vertebral bodies.)

first thoracic vertebra to the third caudal vertebra. Along the entire length of the vertebral column between these two points it attaches to all the vertebral spines.

To expose the spinal cord remove the vertebral laminae from the atlas to the sacrum. Observe the spinal cord with its dural covering. Between the dura of the spinal cord and the periosteum of the vertebrae is the **epidural space,** which contains loose connective tissue and fat and blood vessels.

The spinal cord is divided into segments. A group of dorsal and ventral rootlets leave each spinal cord segment on each side and combine respectively to form the dorsal and ventral roots. These join beyond the spinal ganglion to form the **spinal nerve** at the level of the intervertebral foramen. Open a few intervertebral foramina to see the **spinal ganglia.**

There are eight cervical spinal cord segments, thirteen thoracic, seven lumbar, three sacral, and about five caudal. The roots of the first cervical spinal nerve leave the vertebral canal through the lateral vertebral foramen in the arch of the atlas. The second cervical roots leave caudal to the atlas. The cervical roots of segments 3 through 7 leave the vertebral canal through the intervertebral foramina cranial to the vertebra of the same number. The roots of the eighth cervical segment pass caudal to the seventh (last) cervical vertebra. The roots of all the remaining spinal cord segments pass through the intervertebral foramina caudal to the vertebra of the same number.

In the caudal cervical region over the fifth to seventh cervical vertebrae there is an enlargement of the spinal cord that nearly fills the canal. This is the **cervical intumescence,** whose presence is due to an increase in white matter and cell bodies that are associated with the innervation of the thoracic limb. The intumescence occurs from the sixth cervical segment of the spinal cord through the first thoracic segment. Another enlargement occurs in the midlumbar vertebral region for the innervation of the pelvic limb. The **lumbar intumescence** begins at about the fifth lumbar segment and gradually narrows caudally as the spinal cord comes to an end near the intervertebral space between the sixth and seventh lumbar vertebrae. The narrow end of the parenchyma of the spinal cord is known as the **conus medullaris.** The spinal cord terminates in the **filum terminale,** which is a narrow cord of pia that attaches the conus medullaris to the caudal vertebrae. The **cauda equina** includes the conus medullaris together with the adjacent lumbar and sacral roots which extend caudally in the vertebral canal.

Observe the relationship of the spinal cord segments to the corresponding vertebrae. The only spinal cord segments that are found entirely within their corresponding vertebrae are the last two thoracic and the first two (or occasionally three) lumbar segments. All other spinal cord segments reside in the vertebral canal cranial to the vertebra of the same number (Fig. 196). This is most pronounced in the caudal lumbar and sacrocaudal segments of the cord. In general the three sacral segments lie within the fifth lumbar vertebra and the five caudal segments within the sixth lumbar vertebra.

The nerve roots of the first ten thoracic segments and those caudal to the third lumbar segment are long because of the distance between their origin from the spinal cord and their passage through the intervertebral foramen.

MENINGES

Remove the spinal cord from the vertebral canal by cutting the roots at the intervertebral foramina. Cut the thick, fibrous dura longitudinally along the entire length of the dorsal aspect of the spinal cord. Reflect the dura laterally and examine the dorsal rootlets and their length in different areas of the spinal cord.

The thin arachnoid and pia are apposed to each other and still remain on the spinal cord. On the lateral surface of the spinal cord the pia thickens and forms a longitudinal cord of connective tissue called the **denticulate ligament.** This ligament attaches to the arachnoid and dura laterally midway between the roots of adjacent spinal cord segments. At each intervertebral foramen the dura forms a strong attachment to the foramen, and the subdural and subarachnoid spaces end here.

VESSELS

There is a longitudinal **ventral spinal artery** and one or two dorsal spinal arteries on the spinal cord. They are formed by **spinal branches** of the paired vertebral arteries in the cervical region, intercostal arteries in the thoracic region and lumbar arteries in the lumbar region. The ventral spinal artery is continuous cranially with the basilar artery.

There are vertebral venous plexuses on the floor of the vertebral canal. These were seen previously in the atlanto-occipital region to be continuous cranially with a branch of the sigmoid sinus. The vertebral venous plexuses are in the epidural space. Anastomoses occur between the plexus and the azygos vein and caudal vena cava.

TRANSVERSE SECTIONS

Make transverse sections of the spinal cord at segments C_4, C_8, T_4, T_{12}, L_2, L_6 and S_1. Study these under a dissecting microscope. Compare the shape of the gray matter of these segments and relate this to their areas of innervation.

The gray matter of the spinal cord in cross-section is shaped like a butterfly or like the letter H. It consists primarily of neuronal cell bodies. The dorsal extremity on each side is the **dorsal gray column,** which receives the entering dorsal (sensory) rootlets. The ventral extremity is the **ventral gray column,** which sends axons out in the ventral (motor) rootlets. In the thoracolumbar region the **intermediate gray column** projects laterally from the gray matter midway between the dorsal and ventral gray columns. This contains the cell bodies of the preganglionic sympathetic neurons. In the center of the gray matter of the spinal cord is the **central canal.** This remnant of the embryonic neural tube is continuous rostrally with the fourth ventricle.

The white matter of the spinal cord can be divided into three pairs of funiculi. Dorsally, a shallow, longitudinal groove extends the entire length of the spinal cord. This is the **dorsal median sulcus.** The longitudinal furrow along which the dorsal rootlets enter the cord is the **dorsolateral sulcus.** Between these two sulci is the **dorsal funiculus** of the spinal cord.

Between the dorsolateral sulcus and the line of exit of the ventral rootlets, the **ventrolateral sulcus,** is located the **lateral funiculus.** The **ventral funiculus** is the white matter between the line of exit of the ventral rootlets and the longitudinal groove on the ventral side of the spinal cord, the **ventral median fissure.** In some species the funiculi have been subdivided topographically into specific ascending and descending tracts. Such anatomical information in domestic animals is still incomplete.

VERTEBRAL LIGAMENTS

Examine the vertebral column. On the midline of the floor of the vertebral canal is the **dorsal longitudinal ligament.** This ligament widens where it passes over and attaches to the anulus fibrosus of each intervertebral disk (Fig. 197). It extends the entire length of the vertebral column to the dens of the axis. The **ventral longitudinal ligament** is located on the ventral surface of the vertebral bodies and extends from the axis to the sacrum (Fig. 198).

Notice the thick **transverse ligament of the atlas** (Fig. 199) that spans the ventral arch. It crosses the dorsum of the dens and functions to hold this process against the ventral arch of the atlas. The **alar liga-**

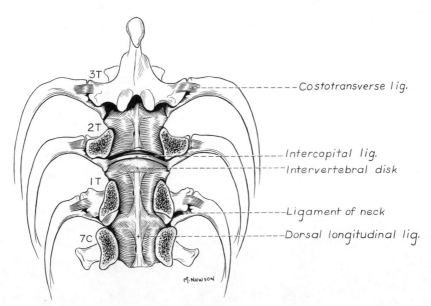

Figure 197. *Ligaments of vertebral column and ribs, dorsal view.*

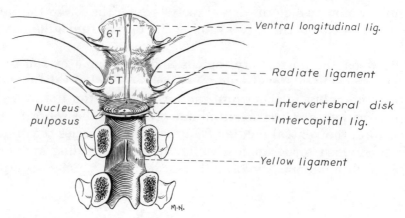

Figure 198. *Ligaments of vertebral column and ribs, ventral view.*

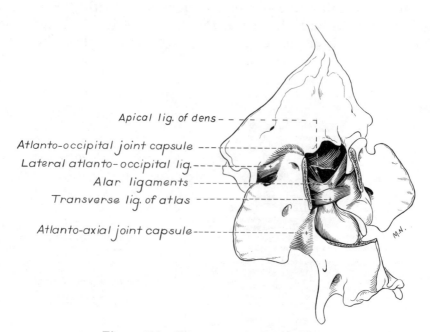

Figure 199. *Ligaments of atlas and axis.*

ments, which connect the dens rostrally with the occipital bone on either side, were cut when the head was removed.

From T_2 to T_{11} a ligament courses transversely across the dorsum of each intervertebral disk and ventral to the dorsal longitudinal ligament. It is attached on each side to the heads of the ribs and holds the heads of the ribs tightly in their articular sockets. These are the intercapital ligaments, which can be demonstrated by cutting transversely completely through a midthoracic intervertebral disk (Figs. 197, 198). Notice the synovial membrane between the ligament and the disk and note the groove that the ligament makes in the disk.

Study the transected intervertebral disk. Notice the laminated, dense, outer anulus fibrosus, which is thickest on the ventral surface. In the center is the gelatinous nucleus pulposus. Cut several disks at various levels of the vertebral column and compare them.

REFERENCES

Aǩaevskiǐ, A. 1968. Anatomiia domashnikh Zȟivotnykh. 608 pp. Moscow, Kolos.

Andersen, A. C. 1970. The Beagle as an Experimental Dog. 616 pp. Iowa State Univ. Press.

Ballard, W. W. 1964. Comparative Anatomy and Embryology. 618 pp. New York, Ronald Press.

Barone, R. 1968. Anatomie Comparée des Mammiferes Domestiques. Vol. 1. Osteologie; 811 pp. Vol. 2. 1066 pp. Arthrologie et Myologie. École Nat. Vet., Lyons.

Baum, H., and O. Zietzschmann. 1936. Handbuch der Anatomie des Hundes. 2nd Ed. 242 pp. Berlin, Paul Parey.

Bolk, L. et al. 1931–38. Handbuch der vergleichenden Anatomie der Wirbeltiere. 6 vols. Berlin, Urban and Schwarzenberg.

Bourdelle, E., and C. Bressou. 1953. Anatomie régionale des Animaux Domestiques. IV. Carnivores: Chien et Chat. 502 pp. Paris, Baillière.

Bradley, O. C., and T. Grahame. 1959. Topographical Anatomy of the Dog. 6th Ed. 332 pp. New York, MacMillan.

Bruni, A. C., and U. Zimmerl. 1950. Anatomia degli Animali Domestici. 2nd Ed. 2 vols., 458 pp. and 736 pp. Milano, Vallardi.

Crouch, J. E. 1969. Text-Atlas of Cat Anatomy. 399 pp. Philadelphia, Lea & Febiger.

Dobberstein, J., and G. Hoffmann. 1954–1961. Lehrbuch der vergleichenden Anatomie der Haustiere. 3 vols. 598 pp. Leipzig, Hirzel.

Ellenberger, W., and H. Baum. 1943. Handbuch der vergleichenden Anatomie der Haustiere. 18th Ed. 1155 pp. Berlin, Springer.

Gardner, E., D. J. Gray and R. O'Rahilly. 1969. Anatomy. A Regional Study of Human Structure. 3rd Ed. 812 pp. Philadelphia, Saunders.

Ghetie, V., E. Pastea and I. Riga. 1954–58. Atlas de Anatomie Comparativa. Vol. I – 771 pp., II – 661 pp. Bucuresti.

Grassé, P.-P. 1967, 1955. Traité de zoologie. Vols. 16 and 17. Mammalia. Paris, Masson et Cie.

Hildebrand, M. 1968. Anatomical Preparations. 100 pp. Berkeley, Univ. Calif. Press.

International Committee on Veterinary Anatomical Nomenclature. 1968. Nomina Anatomica Veterinaria. World Assoc. of Veterinary Anatomists. 146 pp.

Jayne, H. 1898. Mammalian Anatomy. Part 1. The Skeleton of the Cat. 816 pp. Philadelphia, Lippincott.

Kappers, C. U. A., G. C. Huber and E. C. Crosby. 1936. The Comparative Anatomy of the Nervous System of Vertebrates, Including Man. 2 vols. 1845 pp. New York, MacMillan.

Koch, T. 1960–65. Lehrbuch der Veterinär-Anatomie. 3 vols. 1268 pp. Jena, Fischer.

Kovács, G. 1967. Háziállatok anatomiájának atlasza. Munkatársak: G. L. Glósy and E. Guzsal. 444 pp. Budapest, Mezögazdasági.

Mason, M. M. 1959. Bibliography of the Dog. 401 pp. Iowa State Univ. Press.

Miller, M. E., G. C. Christensen and H. E. Evans. 1964. Anatomy of the Dog. 941 pp. Philadelphia, Saunders.

Nickel, R., A. Schummer and E. Seiferle. 1961–1967. Lehrbuch der Anatomie der Haustiere. Bd. 1, 502 pp. and 2, 411 pp. Berlin, Paul Parey.

Papez, J. W. 1929. Comparative Neurology. 518 pp. New York, Crowell.

Popesko, P. 1971. Atlas of Topographical Anatomy of the Domestic Animals. 207 pp. Philadelphia, Saunders.

Romer, A. S. 1970. The Vertebrate Body. 4th Ed. 627 pp. Philadelphia, Saunders.

Schwarze, E. and L. Schröder. 1960–1968. Kompendium der Veterinär-Anatomie. 6 vols. 1708 pp. Jena, Fischer.

Sisson, S. and J. D. Grossman, 1953. Anatomy of the Domestic Animals. 4th Ed. 972 pp. Philadelphia, Saunders.

Singer, M. 1962. The Brain of the Dog in Section. 124 plates. Philadelphia, Saunders.

Stockard, C. R. 1941. The Genetic and Endocrinic Basis for Differences in Form and Behavior. 775 pp. Amer. Anat. Memoirs:*19* Philadelphia, Wistar Inst.

Walker, E. P. et al. 1964. Mammals of the World. 2 vols. text, 1500 pp. 1 vol. bibliography, 769 pp. Baltimore, Johns Hopkins Press.

Walker, W. F. 1970. Vertebrate Dissection. 4th Ed. 402 pp. Philadelphia, Saunders.

INDEX

Folio numbers in *italics* refer to illustrations.

Abdomen, 145–182
  autonomic nervous system of, *168*
  superficial structures of, *146*
Abdominal aorta, *173*
  branches of, *171*
Abdominal cavity, 152
Abdominal mesenteries, *155*
Abdominal muscles, 93–97
Abdominal viscera, 152–182, *158, 159*
  vessels and nerves of, 167–179
Abdominal wall, muscles of, 93
  vessels and nerves of, 145–152
Abducens nerve, 253
Abductor pollicis longus muscle, 38
Accessory carpal bone, 15
Accessory nerve, 103, 256, 265
Acetabular bone, 46
Acetabular lip, 80
Acetabulum, 46, 50
Acoustic meatus, external, 212
  internal, 221
Acromion, *8*
Adductor longus muscle, 63
Adductor magnus et brevis muscle, 63
Adrenal glands, 161
Alar ligaments, 275
Alveolar border, of mandible, 217
Alveoli, 216, 217
Anal canal, 184
Anal region, muscles of, *186*
Anal sac, 184
Anconeal process, 13
Anconeus, 33
Annular digital ligaments, 43
Annulus fibrosus, 85, 277
Ansa subclavia, 121
Antebrachial artery, 137
Antebrachium, arteries of, 136–138
  fascia of, deep, 35
  muscles of, 35–39
Anthelix, 236

Anticlinal vertebra, 88
Antitragus, 236
Anus, 184
Aorta, 115
  abdominal, 171
  thoracic, 117
Aortic arch, branches of, *117*
Aortic hiatus, of diaphragm, 157
Aortic impression, 113
Aortic valve, 126
Appendicular skeleton, 6
Aqueous humor, 243
Arachnoid, 257
Arches
  ischiatic, 47
  of atlas, 86
  vertebral, 85
Arcuate artery, 203
Arcuate line, of ilium, 48
Arm, muscles of, caudal, 30
    cranial, 33
    deep, *27*
    superficial, *25*
  nerves of, 132–136
  superficial structures, *105*
Arterial circle, cranial nerves and, *255*
Arteries
  abdominal, cranial, 146
  antebrachial, deep, 137
  arcuate, 203
  auricular, caudal, 248
  axillary, 127, 128–132
  basilar, 258
  brachial, 127, 132, 137
  brachiocephalic trunk, branches of, *127*
  bronchial, 117
  bronchoesophageal, 117
  carotid, common, 115, 246–248
    branches of, *247*
    internal, 258
  cecal, 174

Arteries (*Continued*)
  celiac, 172
  cerebellar, 258
  cerebral, 258
  cervical, superficial, *117,* 127
  colic, 174
  coronary, 126
  cystic, 172
  deferent, 150
  epigastric, caudal, 145, 193
    cranial, 109
  esophageal, 117
    rami, 173
  ethmoidal, 250
  facial, 247
  femoral, 194, 197
    deep, *193*
  gastric, 173
  gastroduodenal, 173
  gastroepiploic, 173
  genicular, descending, 197
  gluteal, 189
  hepatic, 172
  humeral, circumflex, 130
  iliac, circumflex, 147, 175
  ileal, 175
  ileocolic, *174*
  iliac, 179–182
    external, 192–197
    internal, 189–192
  iliolumbar, 189
  intercostal, 118
  interosseous, 137
  jejunal, 175
  labyrinthine, 258
  lingual, 247
  lumbar, 171
  lumbosacral, *192*
  mandibular alveolar, 250
  maxillary, 248, 250–251
  median, 137
  meningeal, middle, 250
  mesenteric, caudal, 175
    cranial, 174
  of central nervous system, 258–259
  of clitoris, 181
  of forearm and paw, 136–138
  of forelimb, *131*
  of hind paw, *206, 207*
  of pelvic limb, *190*
  of pelvis, female, *180*
    male, *181*
  of penis, 181
  of popliteal region, 197, *198*
  of stifle region, *195, 196*
  of thoracic wall, *106*
  ophthalmic, external, 250
  ovarian, 175
  palatine, 250
  pancreaticoduodenal, 173, 175
  pedal, dorsal, 203
  perineal, ventral, 181
  phrenicoabdominal, 175
  pudendal, external, 145, 148
    internal, 180
  pulmonary trunk, 125
  radial, 137
  rectal, 175
  renal, 175
  sacral, median, 180

Arteries (*Continued*)
  saphenous, 194
    branches of, 195
  spinal, ventral, 258, 274
  splenic, 174
  subclavian, 115, 127
  subscapular, 129
  temporal, deep, 250
    superficial, 248
  testicular, 150, 175
  thoracic, external, 129
    lateral, 129
    thoracodorsal, 130
  thyroid, cranial, 246
  tibial, 197
    cranial, *205*
  ulnar, 137
    collateral, 132
  umbilical, 180
  urogenital, 180
  vertebral, 115
Arytenoid cartilage, 233
Ascending aorta, 115
Atlanto-occipital joint, 239
Atlas, 85–86
  ligaments of, *276*
Atrioventricular valve, 125, 126
Atrium, 124, 126
Auricle, heart, 124–126
Auricular cartilage, 235
Auricular face, of sacrum, 89
Auricular surface, of ilium, 48
Auriculopalpebral nerve, 245
Auriculotemporal nerve, 246, 249
Autonomic nerves, thoracic, *122*
Autonomic nervous system, 118–123
Axilla, 108
  nerves of, *130*
  vessels of, *128*
Axillary artery, 127, 128–132
Axillary lymph node, 109
Axillary nerve, 133
Axis, 86
  ligaments of, *276*
Azygous vein, 114

Basilar artery, 258
Biceps brachii muscle, 33
Biceps femoris muscle, 58
Biliary passages, 170
  ducts, 171
Biventer cervicis muscle, 100
Bladder, urinary, 153, 182
  ligaments of, 152, 183
Blood vessels, great, 116
Bones
  acetabular, 46
  carpal, 14–15
  facial, 211, 213, 216
  hip, 47–50
  hyoid, *213,* 218, *233*
  metacarpal, 16
  metatarsal, *55, 56*
  of forepaw, *14–15*
  of pelvic limb, 46–56
  of skull. See also *Skull.*
    dorsal view, *211*
    lateral view, *212*
    ventral view, *216*

Bones (*Continued*)
  of thoracic limb, 6–17, *16*
  of vertebral column, 84–90
  sesamoid, at metacarpophalangeal joints, 45
    of stifle, 75
  tarsal, 55–56
Brachial artery, 127, 132, 137
Brachial plexus, 121, 128, *129*
Brachialis muscle, 33
Brachiocephalic trunk, 115
  branches of, *117, 127*
Brachiocephalic vein, 114
Brachiocephalicus muscle, 21
Brachium, muscles of, caudal, 30
    cranial, 33
  nerves of, 132–136
Brain, 259–271
Braincase, 210, 212, 214
Brainstem, diencephalon, 270–271
  surface structures of, 262–267
Bronchial arteries, 117
Bronchoesophageal artery, 117
Buccinator muscle, 226
Bulb, of penis, 186
Bulbar conjunctiva, 226
Bulbospongiosus muscle, 186
Bulbus oculi, 242. See also *Eyeball.*
Bulbus glandis, of penis, 187
Bursa, calcaneal, 78
  omental, 153, 166
  ovarian, *164*

Calcaneal bursa, 78
  tendon, 58, 75
Calcaneus, 56
Calvaria, 210
Canal
  alveolar, 214
  anal, 184
  carpal, 43
  for trigeminal nerve, 221
  hypoglossal, 216
  inguinal, 151
  mandibular, 217
  musculotubal, 215
  optic, 214, 221
  pelvic, 47
  petrooccipital, *222, 259*
  vertebral, 85
Canine tooth, 219
Capitulum, of humerus, 10
Cardiac nerves, 121
Cardiac notch, 112
Carotid artery, common, 246–247
  external, 247–248
  internal, 258
Carotid sheath, 26
Carotid sinus, 247
Carpal bones, 14–15
Carpal canal, 43
Carpal joint, capsule of, 46
Carpal pad, 34
Carpus, 14
  ligaments of, 46
Cartilages
  arytenoid, 233
  auricular, 235
  corniculate, 233
  costal, 89
  intersternebral, 90
  laryngeal, *213*, 232, *233*
  scutiform, 227

Cartilages (*Continued*)
  tracheal, 246
Caruncula lacrimalis, 226
Cauda equina, 273
Caudal vertebrae, 89
Caudate nucleus, of cerebrum, 269
Cavities
  abdominal, 152
  cranial, 220
  glenoid, 6, *7*
  nasal, 222–*223*
  of skull, 220–224
  oral, 227–229
  pericardial, 123
  peritoneal, 152
  tympanic, 224
Cecocolic orifice, 160
Cecum, 160
Celiac artery, 172
Celiac ganglia, 169
Central nervous system, 118
Cephalic veins, 103, 134
Cerebellomedullary cistern, 257
Cerebellum, 261–262, 264
Cerebral arterial circle, 258
Cerebral meninges, 257
Cerebrum, surface structures of, 260–261
  telencephalon, 266–270
Cervical intumescence, 273
Cervical nerves, 102–104
  transverse, 103
Cervical structures, 246
Cervical vertebrae, 85–87
Cervix, of uterus, 187
Chambers, of eyeball, 243
Cheek, muscles of, 225
Cheek teeth, 219
Choanae, 217
Chordae tendinae, 125
Choroid, 243
Choroid plexus, 261
Ciliary body, 243
  and zonule, *241*
Ciliary processes, 243
Cingulum, 269
Cisterna chyli, 114
Clavicle, 6
Cleidobrachialis muscle, 24
Cleidocervicalis muscle, 24
Cleidomastoideus muscle, 24
Clitoris, 188
  artery of, 181
Coccygeus muscle, 178
Cochlea, tibial, 54
Cochlear window, 212
Collateral ligaments, 46, 81
Colliculi, of midbrain, 264
Colliculus seminalis, 185
Colon, 160
Column, vertebral, bones, 84–89
  ligaments, 275–277
Columns, gray, of spinal cord, 274
Conchae, nasal, dorsal, 222
    ventral, 223
Condyles, femoral, 53
  humeral, 10
  tibial, 53
Conjunctivae, 226
Conus arteriosus, 125
Conus medullaris, 273
Coracobrachialis muscle, 30
Cornea, 242
Coronary arteries, 126

Coronary groove, 123
Coronary sinus, 125, 126
Coronoid process, of mandible, 217
    of ulna, 13
Corpora quadrigemina, 264
Corpus callosum, 266
Corpus cavernosum penis, 186
Corpus spongiosum penis, 187
Cortex, renal, 162
Costal cartilages, 89
Costal pleura, 109
Costocervical trunk, 116
Cranial cavity, 220
Cranial fossae, 221
Cranial nerves, 251–256, 264
    and arterial circle, 255
Cremaster muscle, 148
Cribriform foramina, 221
Cricoarytenoideus muscles, 234
Cricoid cartilage, 232
Cricothyroid ligament, 232
Cricothyroid muscle, 234
Cruciate ligaments, 81, 84
Crural fascia, 57
Crus, muscles of, 70–79
        caudal, 75
        craniolateral, 71
Crus cerebri, 262
Cuneiform cartilage, 233
Cutaneus trunci, 18
Cystic artery, 172
Cystic duct, 170

Deciduous dentition, 219
Decussation of the pyramids, 265
Deltoid tuberosity, 10
Deltoideus muscle, 28
Dens, of axis, 86
Dental formula, 219
Denticulate ligament, 274
Descending aorta, 115
Diaphragm, 156
Diaphragm, 156, 157
Diencephalon, 262, 263, 270–271
Digastricus muscle, 237
Digits, joints of, 46
    muscles of, extensor, 35
        flexor, 43, 78
    of pelvic limb, blood supply of, 209
        innervation of, 209
    of thoracic limb, arteries of, 144
        innervation of, 144
Dorsum sellae, 221
Ducts, biliary, 170, 171
    cystic, 170
    hepatic, 170
    incisive, 229
    lacrimal, 226
    nasolacrimal, 226
    pancreatic, 160
    parotid, 227, 228
    mandibular, 228
    sublingual, 228
    zygomatic, 236
Ductus choledochus, 171
Ductus deferens, 150, 152
Duodenum, 160
Dura mater, 257

Ear, external, 227, 235–236, 235
    vessels and nerves of, 244
    middle, 224
    inner, 212
Ectoturbinates, 222
Elbow, 46
Endocardium, 125
Endoturbinates, 222
Epaxial muscles, 90, 97–101
Epicardium, 123
Epicondyles, femoral, 53
    humeral, 11
Epididymis, 151
    caudal ligament of, 151
Epigastric artery, caudal, 193
    cranial, 109
Epiglottic cartilage, 232
Epiploic foramen, 166
Epithalamus, 270
Esophageal arteries, 117
Esophageal hiatus, of diaphragm, 157
Esophagus, 246
Ethmoid bone, 223
Ethmoidal arteries, 250
Ethmoidal labyrinth, 223
Extensor carpi radialis muscle, 35
Extensor fossa, of femur, 53
Extensor retinaculum, 35, 71
External acoustic meatus, 212
External obturator muscle, 68
Eye, and related structures, 239–244
Eyeball, 240–241
    muscles of, extrinsic, 240
    nerves of, 252
Eyelids, 226
    third, 226, 239

Face, muscles of, 225–227
Facial artery, 247
Facial bones, 211, 213, 216
Facial nerve, 245–246, 253
Facial vein, 244
Falciform ligament, 152
Falx cerebri, 257
Fascia lata, 57
Fasciae
    antebrachial, deep, 35
    crural, 57, 70
    femoral, 57
    gluteal, superficial, 56
    metatarsal, 70
    of lumbar region, 57
    of neck, superficial, 21
        deep, 26
    of trunk, 90
        deep, 57
    spermatic, 148
    tarsal, 70
    thoracolumbar, 57
    transversalis, 152
Fasciculus cuneatus, 265
Fasciculus gracilis, 265
Femoral arteries, 193, 194, 197
Femoral nerve, 147, 198, 200
Femoral region, deep structures of, 202
Femoral triangle, 62, 194
Femorotibial ligaments, 81
Femur, 50–53, 69
    muscle attachments on, 69

Fibula, *54, 55, 71, 72*
Filium terminale, 273
Fissure, occipitotympanic, 215
Flexor carpi radialis muscle, 40
Flexor carpi ulnaris muscle, 42, 43
Flexor retinaculum, 43, 78
Flexures, colic, 160
  duodenal, 160
Foramen lacerum, 215
Foramen magnum, 217
Foramina
  alar, 214
  cribriform, 221
  epiploic, 166
  intervertebral, 85
  jugular, 221
  mandibular, 217
  mastoid, 217
  of atlas, 86
  of pterygopalatine fossa, 214
  oval, 215
  palatine, 217
  sacral, 89
  stylomastoid, 216
  vertebral, 85
Forearm, arteries of, 136–138
  muscles of, 35, 38, 40
  nerves of, 138–144
Forelimb. See also *Thoracic limb.*
  arteries of, *131*
  muscles of, attachments of, *22, 23*
    major extensors and flexors, *44*
  parts of skeleton of, *16*
  veins of, *140*
Forepaw, arteries of, 136–138
  bones of, *14, 15*
  muscles of, 45
  nerves of, 138–144
  phalanges of, *17*
  superficial structures of, *141, 142*
Fornix, of cerebrum, 268
  of vagina, 188
Fossa ovalis, 125
Fossae
  cranial, 221
  extensor, of femur, 53
  hypophyseal, 221
  intercondylar, femoral, 53
  mandibular, 216
  masseteric, 217
  olecranon, 11
  pterygopalatine, 213
    foramina of, 214
  radial, 11
  subscapular, 7
  temporal, 211
Fovea capitis, of radius, 11
Fovea capitis femoris, 50
Fovea dentis, 86
Foveae, articular, of atlas, 86
  costal, of thoracic vertebrae, 87
Frontal bone, 210
Frontal lobe, of cerebrum, 260
Frontal sinuses, 224

Gall bladder, 170
Ganglia
  abdominal, 169
  celiac, 169

Ganglia (*Continued*)
  celiacomesenteric, 169
  cervical, 120, 121, 256
  cervicothoracic, 121
  geniculate, 253
  mesenteric, caudal, 170
    cranial, 169
  otic, 254
  pterygopalatine, 253
  sympathetic trunk, 119
  vagal, 254
Gastrocnemius muscle, 75
Gastrosplenic ligament, 156
Gemelli muscles, 68
Genicular artery, descending, 197
Geniculate bodies, of thalamus, 271
Geniculate ganglion, 253
Genioglossus muscle, 237
Genitofemoral nerve, 148
Genu, of corpus callosum, 266
Girdle, pelvic, 6, 46
  thoracic, 6
Glands
  adrenal, 161
  lacrimal, 226, 239
  of third eyelid, superficial, 226
  parathyroid, 246
  prostate, 184
  salivary, 228, *229,* 236
  thyroid, 246
  zygomatic, 236
Glans penis, 187
Glenoid cavity, 6, 7
Glossopharyngeal nerve, 254
Glottis, 233
Gluteal arteries, 189
Gluteal fascia, superficial, 56
Gluteal muscles, 67
Gluteal nerves, 201, 204
Gluteal region, *202*
Gracilis muscle, 62
Gray columns, of spinal cord, 274
Great auricular nerve, 102
Greater omentum, 152, 166
Grooves, of heart, 123
Gyri, of brain, 260, 261

Hard palate, 217
Head, 210–256
  bones of, 210–224
  joints of, 239
  muscles of, superficial, *225*
  structures of, 224–256
  veins of, superficial, 244–245
Heart, and pericardium, 123–127
  grooves of, 123
Hepatic artery, 172
Hepatic ducts, 170
Hepatoduodenal ligament, 166
Hilus, of kidney, 161
Hind limb. See *Pelvic limb.*
Hind paw, arteries and nerves of, *206, 207*
  phalanges of, 56
Hip, 79. See also *Hip bone.*
  muscles of, 67–69
Hip bone, 46, 47–50
Hippocampus, 268
Hock, 84
Humerus, 8–11
Humor, aqueous, 243

Hyoglossus muscle, 237
Hyoid apparatus, 218
Hyoid bones, 213, 218, *233*
Hyoid muscles, 238
Hyopharyngeus muscle, 238
Hypaxial muscles, 90, 92–97
Hypogastric nerves, 170
Hypoglossal canal, 216
Hypoglossal nerve, 256, 265
Hypophyseal fossa, 221
Hypothalamus, 262

Ileal arteries, 175
Ileocolic artery, *174*
Ileocolic orifice, 160
Ileum, 160
Iliac arteries, 179–182
    circumflex, deep, 175
    external, 192–197
    internal, 189–192
Iliac crest, 47
Iliac spines, 47
Iliacus muscle, 70
Iliocostalis muscle system, 98
Iliolumbar artery, 189
Iliopsoas muscle, 70
Ilium, 46, 47
Incisive bone, 211
Incisive duct, 229
Incisive papilla, 228
Incisor teeth, 219
Incus, 212
Infraspinatus muscle, 29
Infundibulum, of uterine tube, 165
Inguinal canal, 95, 151
Inguinal ligament, 95
Inguinal lymph nodes, superficial, 146
Inguinal ring, deep, 97, 152
    superficial, 95, 97
Inguinal structures, 94–96
    male, 148
    female, 151
Interatrial septum, 125
Intercapital ligaments, 277
Intercondylar areas, of tibia, 53
Intercondylar fossa, of femur, 53
Intercostal arteries, 118
Intercostal muscles, external, 92
    internal, 93
Intercostal structures, schema, *107*
Internal capsule, of brain, 267
Internal obturator muscle, 67
Interossei muscles, 45
Interosseous arteries, 137
Interosseous ligament, 46
Interosseous membrane, of crus, 84
Intersternebral cartilages, 90
Intervenous tubercle, of heart, 125
Interventricular grooves, 123
Intervertebral discs, 275
Intervertebral foramina, 85
Intestine, 160
Intumescences, of spinal cord, 273
Iris, 243
Ischiatic arch, 47
Ischiatic nerve, 201, 204

Ischiatic notch, greater, 48
    lesser, 49
Ischiatic spine, 49
Ischiatic tuberosity, 48
Ischiocavernosus muscle, 186
Ischium, 46, 48
    ramus of, 49

Jejunal arteries, 175
Jejunum, 160
Joints
    of head, 239
    of pelvic limb, 79–84
    of thoracic limb, 45–46
    of vertebral column, 275–277
    sacroiliac, 79
Jugular foramen, 221
Jugular process, of occipital bone, 213
Jugular vein, external, 103, 244
    internal, 26

Kidneys, 161–164

Labia, 188
Labyrinthine artery, 258
Lacrimal glands, 226, 239
Laminae, of vertebral arch, 85
Large intestine, 160
Laryngeal nerves, 254
Laryngeal ventricle, 233
Laryngopharynx, 232
Larynx, 232–235
    cartilages of, *213, 232, 233*
    muscles of, *233, 234*
Lateral masses, of atlas, 85
Latissimus dorsi muscle, 26
Leg, muscles of, 70–79
Lens, 243
Leptomeninges, 257
Lesser omentum, 166
Levator ani muscle, 178
Levator nasolabialis muscle, 226
Levator palpebrae superioris muscle, 227, 240
Levator veli palatini muscle, 239
Ligaments, acetabular, transverse, 80
    alar, 275
    collateral, stifle, 81
        elbow, 46
    cricothyroid, 232
    cruciate, 81, 82, *84*
    denticulate, 274
    digital, annular, 43
    falciform, 152
    femorotibial, 81
    gastrosplenic, 156
    hepatoduodenal, 166
    humeral, transverse, 46
    inguinal, 95
    intercapital, 277
    interosseous, 46
    meniscal, of stifle, *84*

Ligaments (*Continued*)
  meniscofemoral, 81
  nuchal, 101, 217
  of atlas, transverse, 275
  of axis, *276*
  of bladder, 152
  of carpus, 46
  of epididymis, 151
  of femoral head, 79
  of interphalangeal joints, 36
  of liver, round, 152
  of ovary, 165
  of pelvis, *80, 81*
  of ribs and vertebral column, *275, 276*
  of stifle, *82, 83*
  of testis, proper, 151
  of tibia, *84*
  of urinary bladder, 183
  of uterus, broad, 153
    round, 151, 156
  orbital, 213
  peritoneal, 166
  pulmonary, 111
  sacroiliac, 79
  sacrotuberous, 63, 79
  supraspinous, 101, 271
  vertebral, *275, 276*
Ligamentum arteriosum, 126
Limbus, of cornea, 242
Linea alba, 95
Lingual artery, 247
Lingual frenulum, 228
Lingual muscles, 237
Lingual nerve, 249
Lingual vein, 244
Linguofacial vein, 103
Lips, muscles of, 225
  of femur, 51
Liver, 157, 158
  lobes of, 158
  round ligament of, 152
Lobes, of cerebrum, 260
  of liver, 158
  of lungs, 112
  of pancreas, 160
Longissimus muscle system, 99
Longitudinal fissure, of cerebrum, 260
Longitudinal ligaments, of vertebrae, 275
Longus capitis muscle, 92
Longus colli muscle, 92
Lumbar arteries, 171
Lumbar intumescence, 273
Lumbar nerves, *147*
Lumbar vertebrae, 88
Lumbosacral arteries, *192*
Lumbosacral nerves, *192*
Lumbosacral plexus, 198–205
Lungs, 112–113
  root of, 111
Lymph nodes
  axillary, 109
  cervical, superficial, 103, 127
  inguinal, superficial, 146
  mandibular, 21, 103
  mesenteric, 160
  popliteal, 58
  retropharyngeal, 104, 247
  tracheobronchial, 113
Lymph vessels, thoracic duct, 114
  tracheal ducts, 114
Lyssa, 228

Malleolus, lateral, of fibula, 55
Malleus, 212
Mammae, 17
  abdominal, inguinal, 145
  thoracic, 107, 108
Mandible, 217–218
Mandibular alveolar artery, 250
Mandibular alveolar nerve, 249
Mandibular canal, 217
Mandibular foramen, 217
Mandibular fossa, 216
Mandibular lymph nodes, 103
Mandibular nerve, 248
Mandibular salivary gland, 228
Manubrium, 90
Manus, muscles of, 45
Masseter muscle, 236
Masseteric fossa, 217
Mastoid foramen, 217
Mastoid process, 212
Maxilla, 211
Maxillary artery, 248, 250–251
Maxillary nerve, 253
Maxillary recess, 224
Maxillary vein, 103
Meatus, acoustic, external, 212
  internal, 221
  nasal, 223, 224
Median artery, 137
Median nerve, 134, *135,* 139
Mediastinum, 111
Medulla, brain, 265
  renal, 162
Medullary velum, 266
Membrane, interosseous, of crus, 84
Meningeal artery, middle, 250
Meniscal ligaments, *84*
Menisci, *84*
Meniscofemoral ligament, 81
Mesenteric lymph nodes, 160
Mesenteries, abdominal, *155,* 166
Mesocolon, 166
Mesoductus deferens, 150
Mesoduodenum, 165
Mesometrium, 165
Mesorchium, 150, 151, 175
Mesosalpinx, 165
Mesovarium, 165
Midbrain, 262
Meninges, cerebral, 257
  of spinal cord, 274
Mesencephalon, 262
Metacarpal bones, 16
Metacarpal pad, 34
Metacarpophalangeal joints, *14, 15,* 17
Metacarpus, 16
Metatarsal bones, *55, 56*
Metencephalon, ventral, 264
Molars, 219
Mouth, 227–229
Muscles
  abdominal, oblique, *94*–96, 149
  auricular, 227
  buccinator, 226
  bulbospongiosus, 186
  cremaster, 148
  cricoarytenoideus, 234
  cricothyroid, 234
  digastricus, 237
  epaxial, 90, 97–101
  genioglossus, 237

Muscles (*Continued*)
  hyoglossus, 237
  hyoid, 238
  hyopharyngeus, 238
  hypaxial, 90, 92–97
  iliocostalis, 98
  intercostal, external, 92
    internal, 93
  ischiocavernosus, 186
  laryngeal, *223, 234*
  levator nasolabialis, 226
  lingual, *237*
  longissimus system, 99
  masseter, 236
  of abdominal wall, 93
  of anal region, *186*
  of antebrachium, 35–39
  of arm, 30, 33
  of crus, 70–79
  of eyeball, extrinsic, 240
  of face, 225–227
  of femur, attachments of, *69*
  of forearm, 35, 38
  of forepaw, 45
  of head, superficial, *225*
  of hip, 67–69
  of leg, 70–79
  of mastication, 236–239
  of neck, 92, *93*
    and thorax, superficial, *20*
    deep, *99*
  of pelvic limb, 56–79
    attachments of, *64, 65*
  of penis, 186
  of rump, lateral, 63–67
  of scapula, 28–29
  of scapula, shoulder and arm, deep, *27*
    superficial, *25*
  of thigh, caudal, 58–59
    cranial, 69–70
    medial, 59–63
  of thoracic limb, 17–45
    attachments of, *22, 23*
  of thoracic wall, 92–*93*
  of trunk, 90–101
    epaxial, 90, 97–101
    hypaxial, 90, 92–97
  orbicularis oculi, 226
  orbicularis oris, 225
  palatine, 239
  palatopharyngeus, 238
  papillary, 125
  pectinate, 125
  pectoral, deep, 20
    superficial, 19
  pharyngeal, *237, 238*
  pterygoid, 236
  pterygopharyngeus, 238
  rectococcygeus, 184
  sphincter, of anal canal, 184
  sternocephalicus, 24
  sternohyoideus, 238
  styloglossus, 237
  stylopharyngeus, 238
  temporalis, 236
  thyrohyoideus, 238
  transversospinalis, 100
  vastus, 70
Muscular system, skeletal system and, 6–101
Musculocutaneous nerve, 132, *135*
Musculophrenic artery, 112

Musculotubal canal, 215
Myelencephalon, 265
Mylohyoid nerve, 249

Nasal aperture, 222
Nasal bone, 211
Nasal cavity, 222, 223
Nasal septum, 223
Nasolacrimal ducts, 226
Nasopharynx, 232
Neck, fascia of, 21, *26*
  muscles of, 92, *93*
    deep, *99*
    superficial, *20*
  structures of, 246
    superficial, *103*
  vessels and nerves of, 102–104
Nerves
  abdominal, of viscera, 167–179
    plexuses and ganglia, 169
  abducens, 253
  accessory, 103, 256, 265
  axillary, 133
  auriculopalpebral, 245
  auriculotemporal, 246, 249
  autonomic, 118–123, 167–170
  brachial plexus, 121
  cardiac, 121
  cervical, 102–104
  cranial, 251–256, *251,* 264
  cutaneous, of abdomen, 146
  facial, 245–246, 253
  femoral, 147, 198, *210*
    cutaneous, caudal, 201
  genitofemoral, 148
  glossopharyngeal, 254
  gluteal, 189, 201, *204*
  hypogastric, 170
  hypoglossal, 256, 265
  ischiatic, 201, *204*
  laryngeal, 123, 254
  lingual, 249
  lumbar, *147*
  lumbosacral, *192*
  mandibular, 248
    alveolar, 249
  maxillary, 253
  median, 134, 135, 139
  musculocutaneous, 132, 135
  mylohyoid, 249
  obturator, 198, 200
  oculomotor, 252
  of abdominal wall, 145–152
  of axillary region, *130*
  of external ear, *244*
  of eyeball and its adnexa, *252*
  of forearm and paw, 138–144
  of hindpaw, *206, 207*
  of mandible, *249*
  of neck, 102–104
  of penis, dorsal, 201
  of scapular region and arm, 132–136
  of stifle region, *195, 196*
  of thoracic limb, 127–144
  of thoracic wall, *106*
  olfactory, 251
  ophthalmic, 252
  optic, 251

Nerves (*Continued*)
  palatine, 253
  parasympathetic, 119
  pelvic, 178, *179*
  perineal, 201
  peroneal, 203
  phrenic, 118
  pterygopalatine, 253
  pudendal, 199
  radial, 134, *138*, 139
  ramus communicans, *82*, 119
  rectal, 201
  saphenous, *200*
  sciatic, *192*, 201, 203, *204*
  spinal, 119
    dorsal roots of, *272*
  splanchnic, 120, 169
  subscapular, 132
  suprascapular, 132
  sympathetic trunk, 119, 167
  thoracodorsal, 134
  tibial, 205
  trigeminal, 248–250, 252
    canal for, 221
    spinal tract of, 266
  trochlear, 252
  ulnar, 134, *136*, 139
  vagal, 121, 123, 167, 254
  vagosympathetic trunk, 104, 121
  vertebral, 121
  vestibulocochlear, 254
Nervous coat, of eye, 243
Nervous system, 257–277
  arteries of, 258–259
  autonomic, 118–123, 167–170
    abdominal, *168*
    pelvic, *179*
    thoracic, *122*
  veins of, 259
Neurocranium, 210
Nose, muscles of, 225
Notch, cardiac, 112
  popliteal, 53
  scapular, 8
  ulnar, 11
  vertebral, 85
Nuchal crest, 210, 217
Nuchal ligament, 101, 271
Nucleus pulposus, 85, 277

Obliquus muscles, of eyeball, 240, 241
Obturator muscles, 67, 68
Obturator nerve, 198, 200
Obturator sulcus, 50
Occipital lobe, of cerebrum, 260
Occipitotympanic fissure, 215
Oculomotor nerve, 252
Olecranon, 12
Olecranon fossa, 11
Olfactory nerve, 251
Omental bursa, 153, 166
Omentum, 152, 166
Omobrachial vein, 136
Omotransversarius muscle, 25
Ophthalmic artery, external, 250
Ophthalmic nerve, 252
Optic canal, 214, 221
Optic disc, 243

Optic nerve, 251
Ora serrata, of retina, 243
Oral cavity, 227–229
Orbicularis oculi muscle, 226
Orbicularis oris muscle, 225
Oribt, 213, 214
Oropharynx, 229
Os coxae, 46, 47–50
Os penis, *187*
Otic ganglion, 254
Ovarian artery, 175
Ovary, and ovarian bursa, *164*
  ligaments of, 165

Pachymeninx, 257
Palatine arteries, 250
Palatine fissures, 217
Palatine foramina, 217
Palatine nerves, 253
Palatine tonsil, 232
Palatopharyngeus muscle, 238
Palleopallium, 261
Palpebrae, 226
Palpebral conjunctiva, 226
Pampiniform plexus, 151
Pancreas, 160
  and ducts, *161*
Pancreaticoduodenal arteries, 173, 175
Papilla, incisive, 228
Papillae, of mammae, 17
  of tongue, 227, 228
Papillary muscles, 125
Paraconal interventricular groove, 123
Paranasal sinuses, 224
Parathyroid glands, 246
Parietal lobe, of cerebrum, 260
Parietal pleura, 109
Parietal vaginal tunic, 148
Parotid salivary gland, 21, 228
  ducts of, 227
Pars ciliaris retina, 243
Pars iridica retina, 243
Pars longa glandis, of penis, 187
Pars optica retina, 243
Patella, 52, 69, *85*
Pecten, of pubis, 50
Pectinate muscles, 125
Pectineus muscle, 62
Pectoral muscles, 19, 20
Pedal artery, dorsal, 203
Pedicles, of vertebral arch, 85
Peduncles, cerebellar, 262
Pelvic canal, 47
Pelvic girdle, 6, 46
Pelvic inlet, 47
Pelvic limb, arteries of, *190, 208
  bones of, 6, 46–56
  cross-sections of, 77
  digits of, blood supply and innervation of, 209
  joints of, 79–84
  muscles of, 56–79, *73, 74, 76*
  nerves of, 208
  veins of, *191*
Pelvic nerve, 178
Pelvic outlet, 47
Pelvic plexus, 178
Pelvic symphysis, 79

Pelvic viscera, 182–205
  female, 187, *188*
  male, 184
Pelvis, female, arteries of, *180*
  ligaments of, *80, 81*
  male, arteries of, *181*
  muscle attachments on, *64, 65*
  renal, 162
Penis, 185
  dorsal nerve of, 201
  vessels of, 181, *182*
Pericardial mediastinal pleura, 111
Pericardium, heart and, 123–127
Perineal nerves, 201
Periorbita, 239
Peripheral nervous system, 118
Peritoneal cavity, 152
Peritoneum, 165, *183*
Peroneal nerves, 203
Peroneus longus muscle, 75
Petrosal part, of temporal bone, 212
Phalanges, of forepaw, 17
  of hind paw, 56
Pharyngoesophageal limen, 246
Pharynx, 229–232
  muscles of, *237, 238*
Philtrum, 224
Phrenic nerves, 118
Phrenicoabdominal artery, 175
Pia mater, 257
Pineal body, 271
Platysma, 21, 225
Pleurae, 109
Plexuses, abdominal, 169
  brachial, 121, 128, *129*
  lumbosacral, 198–205
  pampiniform, 151
  pelvic, 178
  venous, vertebral, 259
Plica semilunaris, 226, *239*
Plica vena cava, 111
Pons, 264
  transverse fibers of, 261
Popliteal artery, 197
Popliteal lymph node, 58
Popliteal region, arteries of, *198*
Popliteal surface, 53
Popliteus muscle, 79
Portal vein, tributaries of, *176*
Premolars, 219
Prepuce, 186
  vessels of, *182*
Promontory, of sacrum, 89
Pronator quadratus muscle, 44
Pronator teres muscle, 38
Prostate gland, 184
Psoas major muscle, 70
Pterygoid muscles, 236
Pterygopalatine fossa, 213
Pterygopalatine nerve, 253
Pterygopharyngeus muscle, 238
Pubis, 46, 49, 50
Pudendal artery, external, 145, 148
  internal, 180
Pudendal nerve, 199
Pudendal vein, 148
Pulmonary ligament, 111
Pulmonary pleura, 109
Pulmonary trunk, 125

Pupil, 243
Pylorus, 159
Pyramidal tracts, 264, 265

Quadratus femoris muscle, 68
Quadriceps femoris muscle, 69

Radial artery, 137
Radial carpal bone, 15
Radial fossa, 11
Radial nerve, 134, 138, 139
Radius, 11–12
Rectal nerve, 201
Rectococcygeus muscle, 184
Rectum, 160, 184
Rectus abdominis muscle, 97
Rectus femoris muscle, 69
Rectus muscles of eyeball, 240
Renal arteries, 175
Renal cortex, 162
Renal pelvis, 162
Renal sinus, 162
Retina, 243
Retractor penis muscles, 186
Rhinal sulcus, 261
Rhomboideus muscle, 26
Ribs, 89, *91*
  ligaments of, *275, 276*
Rima glottidis, 233
Rima pudendi, 188
Rump, muscles of, 63–67

Sacral foramina, 89
Sacroiliac joint, 79
Sacrotuberous ligament, 63, 79
Sacrum, 89
Salivary glands, 228, *229*
  mandibular, 21, 228
  parotid, 21, 228
  sublingual, 228
  zygomatic, 236
Saphenous artery, 194, 195
Saphenous nerve, *200*
Sartorius muscle, 59
Scalenus muscle, 92
Scapha, 236
Scapula, 6–8
  muscles of, 25–30
  nerves of region of, 132–136
Sclera, 242
Scrotum, 151
Scutiform cartilage, 227
Sella turcica, 221
Semilunar fibrocartilage, 80
Semimembranosus muscle, 58
Semispinalis capitis muscle, 100
Semitendinosus muscle, 58
Septum, nasal, 223
Septum pellucidum, 268
Serratus dorsalis muscle, 92
Serratus ventralis muscle, 28, 92

Sesamoid bones, of forepaw, *14, 15,* 17
  of metacarpophalangeal joints, 45
  of stifle, 75
Shoulder, *45*
  muscles of, 25–30
Sinus venarum, 124
Sinuses, carotid, 247
  coronary, 125, 126
  frontal, 224
  paranasal, 224
  renal, 162
  venous, of cranial dura mater, 259
Skeleton. See also *Bones.*
  appendicular, 6
  axial, 84, 210
Skull, 210–224
  cavities of, 220–224
Small intestine, 160
Spermatic cord, 150
Spermatic fascia, 148
Sphincter muscles, of anal canal, 184
Spinal artery, ventral, 258, 274
Spinal cord, 271–277
  meninges of, 274
  segments of, 273
  vessels of, 274
Spinal nerves, 119
  dorsal roots of, *272*
Spines, iliac, 47
  ischiatic, 49
  of vertebrae, 85, 88
Splanchnic nerves, 120, 169
Spleen, 156
Splenic artery, 174
Splenic vein, 178
Splenium, of corpus callosum, 266
Splenius muscle, 100
Stapes, 212
Sternebrae, 90
Sternocephalicus muscle, 24
Sternohyoideus muscle, 24, 238
Sternothyroideus muscle, 25
Sternum, 90, *91*
Stifle, 80
  arteries and nerves of, *195, 196*
  ligaments of, *82, 83, 84*
Stomach, 159
Stria habenularis, 270
Styloglossus muscle, 237
Styloid process, of radius, 11, 14
Stylopharyngeus muscle, 238
Subclavian artery, 127
Sublingual salivary gland, 228
Subpelvic tendon, 62
Subscapular artery, 129
Subscapular fossa, 7
Subscapular nerve, 132
Subscapularis muscle, 29
Subsinual interventricular groove, 123
Sulci, of brain, *260*
Supinator muscle, 38
Supracondylar tuberosities, of femur, 53
Suprascapular nerve, 132
Supraspinatus muscle, 29
Supraspinous ligament, 101, 271
Sustentaculum tali, 56
Sylomastoid foramen, 216
Sympathetic trunk, 119, 167
Symphysis pelvis, 79

Talus, 56
Tapetum lucidum, 243
Tarsal bones, 55–56
Tarsal joint, 84
Tarsus, 55
Teeth, 219–220
Telencephalon, 266–270
Temporal artery, deep, caudal, 250
  superficial, 248
Temporal bone, 212
Temporal fossa, 211
Temporal lines, 210
Temporal lobe, of cerebrum, 260
Temporalis muscle, 236
Temporomandibular joint, 239
Tendon, common calcanean, 58, 75
  subpelvic, 62
Tensor fasciae antebrachii muscle, 30
Tensor fasciae latae muscle, 63
Tensor veli palatini muscle, 239
Tentorium cerebelli, 222
  osseum, *222*
Teres major muscle, 30
Teres minor muscle, 29
Testicular artery, 175
Testis, 151
  proper ligament of, 151
Thalamus, 270
Thigh, cross-section of, *77*
  muscles of, caudal, 58–59
    cranial, 69–70
    medial, 59–63
Third eyelid, 226, *239*
Thoracic arteries, 129
  internal, 111
Thoracic duct, 114
Thoracic girdle, 6
Thoracic limb, arteries of, *131,* 143
  bones of, 6–17, *16*
  joints of, 45–46
  muscles of, 17–45
    attachments of, *22, 23*
    extrinsic, and related structures, 19–28
    intrinsic, 28–45
    major extensors and flexors, *44*
  nerve supply of, 143
  veins of, *140*
  vessels and nerves of, 127–144
Thoracic vertebrae, 87–88
Thoracic wall, arteries of, *106*
  muscles of, 92
  nerves of, *106*
Thoracodorsal artery, 130
Thoracodorsal nerve, 134
Thoracolumbar fascia, 57
Thorax, 105–118
  autonomic nerves of, *122*
  muscles of, *93*
    superficial, *20*
  superficial structures of, 17–19, *108*
Thymus, 111
Thyroid artery, cranial, 246
Thyroid cartilage, 232
Thyroid gland, 246
Thyroarytenoideus muscle, 234
Thyrohyoideus muscle, 238
Tibia, 53–55
  menisci and ligaments of, *84*
  muscle attachments on, *71, 72*

Tibial artery, 197, 205
Tibial cochlea, 54
Tibial nerve, 205
Tibial tuberosity, 53
Tibiofibular joints, 81
Tongue, 227, *237*
Tonsil, palatine, 232
Trabecula septomarginalis, 125
Trabeculae carneae, 125
Trachea, 246
Tracheal cartilages, 246
Tracheal ducts, 114
Tracheobronchial lymph nodes, 113
Tragus, 236
Transversalis fascia, 152
Transversospinalis muscle system, 100
Transversus abdominis muscle, 97
Trapezius muscle, 26
Trapezoid body, 264
Triangle, femoral, 62, 194
Triceps brachii muscle, 31
Tricipital line, of humerus, 10
Trigeminal nerve, 248–250, 252
　canal for, 221
　spinal tract of, 266
Trigone, of bladder, 184
Trochanteric fossa, 50
Trochanters, of femur, 50, 51
Trochlea, femoral, 52
　humeral, 10
　of eye, 241, 242
　radial, 11
Trochlear nerve, 252
Trochlear notch, 13
Trunk, fasciae of, 90
　muscles of, 90–101
　　epaxial, 90, 97–101
　　hypaxial, 90, 92–97
Tuber calcanei, 56
Tuber sacrale, 48
Tubercle, intervenous, 125
　of humerus, 10
Tuberosity, ischiatic, 48
　radial, 11
　supracondylar, of femur, 53
　tibial, 53
Tympanic bulla, 212
Tympanic cavity, 224

Ulna, 12–14
Ulnar artery, 137
Ulnar carpal bone, 15
Ulnar nerve, 134, 136, 139
Ulnar notch, 11
Ulnar tuberosity, 13
Ulnaris lateralis muscle, 38
Umbilical artery, 180
Ungual crest, 17
Ungual process, 17
Ureter, 162
Urethra, female, 188, 189
　male, 185
Urogenital artery, 180
Urogenital system
　bladder, 153–154, 182–184
　female, 153–156, 164–165, 187–189
　kidney, 161–164
　male, 184–187

Uterine tube, 165
Uterus, 153
　cervix of, 187
　ligament of, broad, 153
　　round, 151, 156
Uvea, 243

Vagina, 188
Vaginal process, 95, 150, 151
Vaginal ring, 152
Vaginal tunic, 95, 148, *150, 152*
Vagosympathetic trunk, 104, 121
Vagus nerve, 121, 123, 254
Valves, atrioventricular, 125, 126
Veins, *177*
　azygous, 114
　brachiocephalic, 114
　cephalic, 103, 134
　deferent, 150
　facial, 244
　gastroduodenal, 178
　great cardiac, 126
　jugular, external, 103, 244
　　internal, 26
　lingual, 244
　linguofacial, 103
　maxillary, 103
　mesenteric, 178
　of forelimb, *140*
　of head, superficial, 244–245
　of nervous system, 259
　of pelvic limb, *191*
　omobrachial, 136
　pampiniform plexus of, 151
　portal, 176
　pudendal, external, 148
　splenic, 178
　testicular, 150
　vena cava, 125
　vertebral, plexuses of, 259
Vena cava, caudal, 125, *171*
　cranial, 114, 125
Ventricle, laryngeal, 233
　lateral, of cerebrum, 268
　of heart, *124*
Vertebrae, caudal, 89
　cervical, 85–87
　formula in dog, 84
　lumbar, 88
　thoracic, 87–88
Vertebral arch, 85
Vertebral artery, 115
Vertebral canal, 85
Vertebral column, 84–90
　ligaments of, *275, 276*
Vertebral foramen, 85
Vertebral nerve, 121
Vertebral notches, 85
Vertebral venous plexuses, 259
Vessels. See also *Arteries* and *Veins.*
　of abdominal viscera, 167–179
　of abdominal wall, 145–152
　of axillary region, *128*
　of external ear, *244*
　of pelvic limb, 189
　of pelvic region, *179*
　of spinal cord, 274
　of thoracic limb, 127

Vestibule, of oral cavity, 227
   of vagina, *188*
Vestibulocochlear nerve, 254
Viscera, abdominal, 152–182, *158, 159*
    vessels and nerves of, 167–179
   of female dog, *154*
   of male dog, *154*
   pelvic, 182–205
    female, *188*
   thoracic, 112, *114, 115*
Visceral vaginal tunic, 150
Vitreous body, 243
Vitreous chamber, 243
Vulva, 188

Wall, abdominal, muscles of, 93
   thoracic, muscles of, 92

Xiphoid process, 90

Zonule, 243
   ciliary body and, *241*
Zygomatic arch, 212, 213
Zygomatic salivary gland, 236